CONQUERING

R

BASICS

By Brice Richard

A GOSHEN PUBLISHERS BOOK   VIRGINIA

Conquering R Basics

ISBN: 978-1-7342639-6-1

Copyright ©2021 Brice Richard

Library of Congress Cataloging-in-Publication Data

Published in 2021 by:

GOSHEN PUBLISHERS LLC
P.O. Box 1562
Stephens City, Virginia, USA

www.GoshenPublishers.com

Our books may be purchased in bulk for promotional, educational, or business use. For inquiries please contact the publisher via email: Agents@GoshenPublishers.com.

First Edition 2021

Cover designed by Goshen Publishers LLC

Printed in the United States of America

Dedication

This book is dedicated to you, the reader,
on whose journey this study will position you
closer to a truth, wherever it may lead

Contents

Chapter 1: Introduction 1

Chapter 2: First Principles of R 9

Chapter 3: Practical R Functions for the Beginner 19

Chapter 4: Unlocking the Data Narrative 53

Chapter 5: Application of Color in Plot Development 93

Chapter 6: Video Library of Common Tasks in R 103

PREFACE

This book attempts to tackle a colossal challenge. It proposes to define what is otherwise an ambiguous interpretation of a basic skill set underlying a global technology. That technology is R. Authors proclaiming to have captured a basic R skill set have written books upwards of 350 pages[1] in length describing it. It is difficult to believe that a basic skill set comprising any technology today would require such lengthy prose. That proposition seems to be both excessive and exaggerated. While it is evident that the scope of R technology is indeed broad, there remains a core set of principles underlying concise prose that can be used to adequately describe a basic technological skill set. Technical explanations characteristic of such prose can be strategically constructed in a way that is both targeted and informative. However, a basic technological skill set should not only be controlled by an incisive periphery around which structure, context, and length are formulated, but it should also be thematically interconnected.

If a periphery with such fine distinctions is required to define a basic technological skill set, the question then becomes, *what lies beneath the periphery?*

To answer this question, a brief but personal explanation of recent events is in order. Before this book was written, I spent five years developing a knowledge-based tool called R³ PEGUSYS. *PEGUSYS* is an acronym for ***Pe**rformative **Gu**idance **Sys**tem*. *Performative guidance* is defined as a capabilities-driven strategy used to identify, interpret, and implement the highest level of performing capabilities in a technology unique to a set of specifically defined characteristics. In the case of the PEGUSYS tool, the technology is R, and the specifically defined characteristics of that technology are *software engineering*, *data science proper*, and *machine learning*. Combining *performative guidance* principles within a software application generates a highly targeted, controlled system. The genesis of this techno-philosophical concoction of ideas resulted in R³ PEGUSYS.

The utility of such a system is believed to be immeasurable. On any given data project or problem, current or emerging, PEGUSYS can be used to optimally generate a statistically significant data narrative, statistical model, plot, or dataset. Consequently, the tool is not limited to the creation of these artifacts. It can also be used to construct complex data process models. Optimality in the performance of the tool is defined not only by an increase in productivity, but also by its creative flexibility to connect the right solution to the right problem. More specifically, *performative guidance* approaches problem solving from an unorthodox perspective.

The classic problem-solution model first identifies a problem after which a search for a solution is initiated.

---

[1] For examples supporting this statement, see the following books: *R for Dummies, 2nd Ed* (2015) by Andrie de Vries & Joris Meys is 400 pages, an examination of R graphics called *R Graphics Cookbook* (2013) by Winston Chang is 383 pages, and *Learning R* (2013) by Richard Cotton comes in at around 364 pages. While Chang's book presumes the reader to have "at least a basic understanding of R," it operates from the perspective of simple examples referred to as "recipes." These "recipes" are designed to convert R data into R plots. It is not clear whether this book was designed to target a basic, intermediate, or advanced scope of instruction. Perhaps its scope is an eclectic combination of possibilities.

Performative guidance approaches the same challenge by defining and maintaining an active collection of solutions. The focus is built more on acquiring a deeper understanding of a collection of solutions in the abstract than on the particulars of a problem per se. Endemic to *performative guidance* thinking is the cliché, "I have 1,000 solutions looking for a problem." Inverting the logic commonly associated with problem-solution thinking not only transforms the problem-solution relationship, it redefines the conception of the problem-solution model itself. What makes this method of thinking unique is that the elegance of a solution may solve the particulars of a problem in ways not previously imagined. It is believed that this shift in thinking provides an inherent edge in problem solving.

A compelling use case for *performative guidance* can be found in how it was applied in developing a solution to the problem of defining an *R basic skill set*.

The concept of *performative guidance* originated from having observed significant growth rates in R technology over a short period of time. For example, near the end of 2014, there were a total of 5,428 R packages made available to the public, stored in an online repository called CRAN, the *Comprehensive R Archive Network*. On November 3, 2015, R packages totaled 7,938. On July 6, 2017, that number increased again to 10,990 R packages. As of this writing in May 2021, there were 17,617 active R packages available on CRAN. This represents a 3x increase in package growth in only 7 years. With the sheer number of packages and the accretion of functional options available in R, it became clear that a robust strategy for identifying and applying R functional capabilities to data challenges was needed.

To technically apply *performative guidance* as a capabilities-driven strategy, a systems engineering approach was used to model the underlying framework that ultimately became the PEGUSYS system. The basic design consisted of three distinctive but interrelated sub-component parts. The result was a *performative guidance* system that generated solutions used to develop the content for this book. The idea was to use data analytics to write about data analytics.

The *R basic skill set* proposed in this book comprises the following targeted elements:

- (93) indispensable R functions are identified and explained, all of which can be used in a vast array of data projects.
- The concept of the Data Narrative is introduced and explained.
- Data is programmatically connected with commonly used plot visualizations, explained with specificity and supporting context.
- The application of color in plot development is introduced and explained.

The principal methodologies used in this book to explain an *R basic skill set* are as follows:

- Six video tutorials are integrated into this book (Chapter 6), combining both a reading and an audio-visual experience to facilitate the learning process.
- R technology is explained through the concept of First Principles, which is believed to ease the anxiety commonly associated with learning. Deploying the learning model of First Principles reduces the gap between the ability to comprehend new concepts against the ability to effectively apply them.
- Learning is amplified through a "stair-step" method, in which concepts are first introduced, then are built one upon the other through increasingly advanced concepts, all of which are intricately connected.

- Distinctions in learning are made by changing the mechanism used to foster the learning experience. For example, the reading experience is connected to learning by associating R code directly with functions. Conversely, the audio-visual experiences provided are connected to learning by associating R code with specifically defined tasks built on multi-functional frameworks.

Conquering R Basics is 106 pages in length, of which six are blank to accommodate your notes and personal comments. One Notes page can be found at the end of each chapter. You are encouraged to create and use these Note pages to help guide and reinforce learning.

It is almost axiomatic that for every technical book written, inadvertent errors, omissions, and mistakes will be made in the printed manuscript. Referred to as errata, errors identified within the book's manuscript subsequent to its publication will be captured and addressed on the publisher's website which can be found at:

https://www.goshenpublishers.com/brice-richard

If you find an error that needs to be corrected, please go to the following link or scan the green QR code to report it:

https://www.goshenpublishers.com/contact-us

You can visit my Goshen Publisher's webpage by scanning the dark blue QR code or go to:

https://www.goshenpublishers.com/brice-richard

In conclusion, this book could not have been written without making a few assumptions about the expected reading experience. Consequently, this book is crafted in a way that presumes the following:

- You are able to successfully download, install, and open the R software applications consistent with the instructions provided in Chapter 3.
- You are able to successfully configure the RStudio software to be visually consistent with the RStudio screenshots provided in Chapter 3.
- The book will be read in sequential order from the Preface to Chapter 6; for the first-time reader, it is not designed to be read any other way.

If these assumptions are correct, you will benefit from the best possible learning experience this book has to offer.

This book was made possible by the extraordinary publishing team at Goshen Publishers without whose assistance and tireless contributions to this project would have ensured that this work would have never been published.

With sincere gratitude and thanks to Goshen Publishers.

BR - June 2021

CHAPTER 1
INTRODUCTION

R is one of the most extraordinary tools ever developed in the history of technology. With R you can convert data into graphical visualizations, build complete datasets, and generate comprehensive data profile reports. Each of these tasks can be achieved using one line of code. Even more compelling is that every mathematical framework, scientific formula, and statistical model ever proven will eventually be accessible through this technology. If it is not available today it will be there in the future – most likely sooner than later.

These claims, while seemingly hyperbolic, reflect the rallying cry from a global collaboration of developers and technologists striving to attain a common goal. What emerged from that global collaboration was the development of a new technology in which the mathematical and statistical standards of computation were successfully formalized, reaching critical mass in late 2015. That technology was R. It was a stunning achievement two decades in the making.

R is not a hackneyed software application developed by a leaden organization. It is an ongoing collaboration in which developers, technologists, and academicians from institutions around the world promote and share scientific knowledge. Nearly every domain of scientific knowledge is represented in some form within R technology. An extraordinary by-product of this global collaboration can be found in the development of new data frameworks and coding structures that have emerged from domain cross-pollination.

For example, the concept of *Adaptive Resource Management* or *ARM* moves to develop strategies used in managing natural resources under extreme conditions of uncertainty. Principles formalized under ARM originated in the field of Ecology, but were also heavily influenced by other domains including Physics and Systems Analysis. On closer examination, ARM can also be applied to challenges found in machine learning. In fact, these same principles were used to develop a structural framework in R called *AMModels*, the objective of which is, "to codify knowledge in the form of models and to store it."[2] Models of all types, including those supporting machine learning can be used, organized, and stored within the framework to manage ecological and conservation-based projects. Conversely, the principles underlying ARM can be used to better manage the involutions unique to a machine learning project. Through R technology, an improbable symbiosis emerged, galvanized by the domain cross-pollination of Artificial Intelligence and Ecology.

In another example, the functional capabilities defined by a regular expression were re-contextualized for use in Data Science. In the realm of Software Engineering, one of the most challenging aspects of developing an algorithm involves the successful application of regular expressions. A regular expression programmatically applies a specific sequence of characters that

---

[2]Donovan TM, Katz JE (2018) AMModels: An R package for storing models, data, and metadata to facilitate adaptive management. PLoS ONE 13(2): e0188966.
https://doi.org/10.1371/journal.pone.0188966

are used to define a particular search pattern. Once constructed and tested, the search pattern is then integrated into a larger algorithm that executes the regular expression. Upon execution, the regular expression finds the data string for which it was designed and then additional coding algorithms are executed against the result.

The example provided below defines a typical regular expression, the intent of which is to search for and return all dates found within a body of text:

$$(?<Month>\d\{1,2\})/(?<Day>\d\{1,2\})/(?<Year>(?:\d\{4\}|\d\{2\}))$$

There are two fundamental problems underlying the design and use of regular expressions: 1) the process by which regular expressions are constructed, and 2) testing protocols. Regular expressions are inherently difficult to design correctly, which invariably leads to inaccurate search results. These oversights are caused by an incomplete or incorrectly constructed regular expression. For example, executing the regular expression cited above against a sizeable body of text would return a paucity of dates. More specifically, *no date* matching any of the formats listed below would be returned because the regular expression was not correctly encoded to recognize them:

Date Format	Date Example
yyyy-mm-dd	1976-10-08
dd-mmm-yy	03-Mar-78
mmm d, yyyy	May 8, 2029
m.d.yyyy	7.8.1954

These results expose the second problem associated with regular expressions – testing protocols. If the testing of a regular expression is not comprehensive in scope, it will fail. In the example cited, if all date format variations are not tested against the regular expression, errors and oversights in the expression's syntax construction cannot be identified. While this assessment seems obvious, applying a solution leads to another insidious challenge unique to regular expressions. How do you define all the variations of a particular result? In the case of the cited example, how do you know in what format a date will be manifested in text? It is very difficult to test something against which the rules are not clearly defined, or for that matter, known.

A novel solution to the challenge of regular expressions can be found in R technology's use of *Natural Language Processing*, or *NLP*. Natural Language Processing is a sub-component of Artificial Intelligence. NLP uses advanced computational algorithms to process natural language. An NLP-based solution to the problem posed by the application of regular expressions is outlined in the following steps:

- Disaggregate the entire body of text into words that are grouped by their frequency of occurrence.
- Remove words that are not semantically significant to the analysis. These words are commonly referred to as Stopwords.
- Organize the word listing by sorting it.

In this example, NLP provides a much more effective solution in solving the problem than a regular expression. Each of the steps in the NLP solution can be programmatically executed within algorithms made available through various NLP-based packages in R. By developing a solution that applies Natural Language Processing, every uniquely formatted date contained within a body of text can be identified. If enough text is analyzed, a body of knowledge can be created that identifies a substantial subset of date formats. The results can then be used to develop a standard set of formatting rules against which a variety of date formats can be correctly extracted from future data content.

Another approach used to re-contextualize regular expressions can be found in R's *RVerbalExpressions* package.[3] This package integrates both functional and syntax decomposition to facilitate the construction of regular expressions. Functional decomposition occurs by categorizing chunks of functionality through the creation of a regular expression object. Syntax decomposition can then be applied to build complex regular expressions by "chaining" together these functional chunks. An example of programmatically "chunking" and "chaining" a regular expression is provided below:

This code-syntax approach, driven by the RVerbalExpressions R package, returns a logical value of TRUE or FALSE for each data point in a target data field where the words "house" and "gun" are found.

CODE SYNTAX LEGEND

Where rx	=	Regular Expression Object
Where grepl	=	Global Regular Expression Applying a Logical Result (TRUE or FALSE)
Where DS$FLD	=	Dataset Name and Data Field Name to which the Regular Expression is Applied
Where %>%	=	Code Syntax Chaining Separator

```
rx() %>%
    rx_find("house") %>%
    rx_something() %>%
    rx_find("gun") %>%
    grepl(DS$FLD)
```

While these examples demonstrate advanced concepts in the application of Data Science, they are included to show the capacity and range of possibilities provided by R technology. One solution proposes applying R technology to re-contextualize a functional capability traditionally found in Software Engineering. In a more organic solution, domain cross-pollination reveals a more effective way to solve the problem. By re-contextualizing the dimensions of a solution relative to the problem space, the domain cross-pollination of Software Engineering and Natural Language Processing converged. Separately defined, these domains could not be more different.

Understanding the cross-pollination of functional capabilities emanating from various domains of scientific knowledge can be used to explore new and deeper ways of thinking. After all, the more dimensions of a solution space that can be critically explored, the more dynamic an understanding of the problem space will be. To that end, R is a facilitating technology without which domain

---

[3]Tyler Littlefield (2019). RVerbalExpressions: Create Regular Expressions Easily. R package version 0.1.0. https://CRAN.R-project.org/package=RVerbalExpressions

cross-pollination of collateralized principles and methods, immured within the framework of mathematics, would be missing. In short, R technology redefines the solution space.

R epitomizes the flexibility of a mature technology designed to merge capability with applicability. Its capability, unleashed, will challenge if not completely redefine how you think about data. It has a tendency to capture the imagination of those who are willing to try something new. Of course, there is always the risk that R will turn assumptive thinking completely upside down. Data, as it is applied in R are protean in nature often cloaked in a form factor worn as a means of description within a mechanism of discovery.

Conquering R Basics is designed for the consummate beginner – someone who has little to no knowledge of R. However, it is not just another R "basic guide" book. By comparison to other introductory books on the topic, it is quite different. This work pursues an unorthodox approach in presenting a framework for learning R. Perhaps the most notable feature characterized by this approach can be found in the bifurcation between strategy and focus. The book is constructed in a way that exalts strategy over convention. For example, one of its guiding principles is to provide the technical skills necessary to effectively convert data into a graphical representation, commonly referred to in R as a plot. From the perspective of focus, the work is structured in a way that streamlines the raft of possibilities to something both consumable and practical, building a foundation of prescriptive knowledge one step at a time.

The content for this book is largely derived from the results provided by a knowledge system I built called R^3 PEGUSYS. The goal of this system was to capture, organize, and document the most compelling functional capabilities in the R enterprise framework. This enterprise framework, more formally referred to as the Comprehensive R Archive Network (CRAN), is an online repository of functional capabilities consisting (at the time of this writing) of over 17,000 packages. This knowledge system targets exceptional functional capabilities in domains of technology specific to Software Engineering, Data Science, and Machine Learning. From information contained within PEGUSYS, a frequency analysis was conducted. The analysis evaluated the frequency occurrence of R-based functions referenced in over 170 articles I wrote on various topics related to R. A group of commonly referenced functions from these articles were then identified, forming the basis of this book. Not only do these functions play a critical role in understanding the fundamental principles of R, but are inherently flexible in utility across a wide range of tasks and projects.

What follows are 20 introductory statements made about R. These statements provide supplemental details supporting the technology:

- R is a world class desktop-based software application used for statistical computing and graphics that is supported by the *R Foundation for Statistical Computing.*
- R capabilities are fundamentally defined by what are termed "packages." Each R package consists of a bundled set of capabilities called functions.
- The Comprehensive R Archive Network (CRAN) is an online repository containing thousands of R packages that can be downloaded and used in an R-based application.
- An alternative technology to R is a programming language referred to as Python.
- R has a windows shell that can be installed called RStudio that facilitates both a better understanding and navigation of the technology.

- R, along with the software that supports it, called RStudio, is open source meaning it's free to download and use.
- R technology offers some of the most diverse capabilities of any software development application in the world.
- Since 2018, on average, five R packages a day from around the world are either updated or newly added to CRAN.
- It is possible to produce extraordinary plots in R using three lines of code or less.
- R was conceptualized and initially designed by Statisticians not Software Engineers.
- It is possible to integrate R functionality into third-party applications to create enhanced capabilities that applications, operating independently, could not achieve.
- R can be used to support a variety of domains including but not limited to Biology, Bioinformatics, Conservation, Cybersecurity, Data Simulation Modeling, Education, Genetics, Geospatial Technology, Neural Networks, Machine Learning, Software Engineering, and Statistical Graphing & Modeling.
- R offers a rich user interface with supporting functionality that can be used as a standard data mining process model referred to as "rattle."
- R is a programmatic super calculator that is both easy to use and flexible, offering extended capabilities in vector mathematics.
- R can generate both interactive and static plots.
- R functions typically generate a significant amount of functionality due to the scope and type of mathematical frameworks underlying them.
- R supports extensive documentation that is clear, concise, consistent, and ubiquitous. For example, every R package is specifically documented.
- R can be used to generate reproducible, high-quality data analyses.
- R is an interpreted language that is remarkably flexible. Its primary capability lies in the scope and utility of its functions which can be represented in the form of a value, a vector, an object, or a data frame.
- Updating R is both easy to learn and do.

The greatest capability immured within the fabric of R is that it abstracts complexity away from the user. This means, for example, that it does not require an academic degree in mathematics or statistics to effectively use. Do you need to know all the architectural, electrical, or plumbing intricacies of a house to live in it? Do you need to know the details of an aircraft's construction process to fly as a passenger? While R is built on the rigor of complex mathematical frameworks, it is constructed in a way that allows anyone with a curious interest in the technology to successfully learn and use.

To showcase the power of complexity abstraction in R, a simple example is provided below:

```
x = 1:1000000
y = x + 5
```

What is the value of y? If you believe the answer is 1,000,005 you would be wrong! To understand the solution to this problem you must first understand what the value of x represents. In this example, x represents what is called a vector of 1,000,000 values. A vector is a technical term used in R to denote a series of elements. In this example, these elements are numeric in form. So a 1 followed by a colon and then 1,000,000 executes an instruction to convert the object of x to a vector of 1,000,000 sequentially ordered numeric values starting with 1. To review the first few values of x, applying the code below returns the following:

```
head(x)
[1] 1 2 3 4 5 6
```

Don't be confused by the characters [1]. The [1] denotes that the return values of x are represented by what is called the element number. In this example, the element number represents the values 1 through 6.

The next line then adds 5 to each value in the x vector, returning y as a vector result. So, if we were to review the first few values of y, applying the code below returns the following:

```
head(y)
[1] 6 7 8 9 10 11
```

What this means is that within only two lines of R code, we were able not only to create a vector of 1,000,000 sequentially ordered numeric values, but then add 5 to each of these sequentially ordered values beginning with 1. The output is then converted to a vector called y. While the computation in this example is simple in design, it is extensive in scope. Not only were the results returned almost instantaneously, but no knowledge of the mathematics of addition were required. One million computations were successfully performed with nothing more than an expression of the concept.

This example teaches three important lessons regarding the intersection of technology and mathematics. First, automation can radically transform the relationship between technology and any domain with which it converges. Second, the convergence of technology and mathematics paves the way to substantially reduce complexity. Third, computational functionality redefines mathematics by leveraging automation. A prime example of this phenomenon can be found in *vector mathematics*. The code example provided earlier uses vector mathematics to solve for y. By applying a classic expression such as x + 5, mathematical functionality is transformed by performing computations en masse through automation.

Changes consistent with a perennial trend in the growth of technological innovation over the last two decades have created a permanent transformation in the educational landscape. Knowledge no longer lives in scarcity but thrives in abundance as a form of knowledge democratization. Consequently, the democratization of knowledge is no longer a cliché, it's a reality. R could not provide a better example of a technology that epitomizes the democratization of knowledge. This democratization, manifested in the form of an ongoing global collaboration in which the accessibility, availability, utility, and scalability of R are fully supported, offers an ideal environment in which to pursue knowledge. To that end, time spent learning R denotes a remarkable investment in the development of a more remarkable skill set. This book will argue that it is an investment worth pursuing. More importantly, the ability to learn, relearn, and unlearn is proving to be the indispensable business card of success for the twenty-first century. That which is invested *will* be tested. The more time you invest in the pursuit of knowledge, the more interesting the results will be.

<u>NOTES</u>

CHAPTER 2
FIRST PRINCIPLES OF R

Learning R is a baptism in fire.

Learning, by its very nature, is a vulnerable undertaking. For many, it is a particularly risky venture. After all, the concept of learning anything meaningful is time-consuming, emotionally disorienting, and requires a sufficient amount of patience and focus to succeed. Nothing is more demoralizing than experiencing the emotional discontent of something thought to have been learned only to discover the inability to correctly apply it.

To help mitigate this possibility, this chapter will focus on a critical part of the learning process.

Before R code can be applied with any reasonable degree of skill and confidence, R as a concept must be understood from a position of First Principles. Sometimes referred to as Core Principles, First Principles define a set of concepts so fundamental that they cannot be further reduced. First Principles provide not only the context, but also a strategic approach to learning. Understanding the ability to reason by First Principles reduces the learning curve by easing the tensions between the knowledge acquired from that which can actually be applied. Reducing these tensions reinforces the guiding principle of the learning process. After all, learning is a transformative experience. The fundamental value proposition in learning is to be able to correctly apply what you've learned. To that end, reasoning by First Principles provides a prescriptive roadmap for learning.

In an introduction to R, there are five First Principles. The First Principles outlined in this chapter are not ranked in any particular order. Rather, each principle should be given equal consideration and import as a basis for building a core understanding of R technology. Each principle describes a different dimension of the technology. Collectively, these dimensions formulate a unified understanding within which the technology operates. To minimize the propensity for principle enumeration and to maximize the learning experience, an image icon has been added to each First Principle. Each icon is designed to reflect a corresponding principle theme.

FIRST PRINCIPLE:

Data in R is generally represented as either a quantitative or a qualitative measure.

This means that data can be represented as either a numeric value or as a categorical classification. Examples are provided below:

Quantitative
$6.72
.55
.0045871
319

Qualitative
Fruit
State
Species
Model

Having seven oranges is a quantitative measure. Knowing that the oranges belong to a category of fruit is a qualitative distinction.

The cognitive means of observation should be developed to facilitate the recognition of these data distinctions. When presented with data never before seen, one should be able to easily discern, with specificity, the nature of data by its quantitative or qualitative terms. Viewing data through the lens of this principle promotes a contextual understanding of R. In addition, the ability to make this fundamental distinction is critical when encoding R algorithms. Some R algorithms require quantitative input while others only accept data in a qualitative form. In other cases, a combination of both quantitative and qualitative data are used. Familiarity with these data distinctions plays a key role when R coding algorithms are used to plot data.

Being able to correctly identify the distinction between quantitative and qualitative measures of data is one thing, but how does R technology define this distinction? What is the connection between the utility of data and its computational transformation? How is the interaction between the user and R defined?

The answer, thematically applicable to each of these questions, can be found within the concept of a *data type*. Commonly used in both software engineering and database development theory, a data type defines data by a primary attribute unique to that data. Data types are parameterized in software through the use of code, and in database technologies as a configurable option. Dates, binary data, alphanumeric characters called strings, and numeric values disaggregated by range typify the concept of a data type. A data type can also be defined as unique or special if it provides an exceptional means by which to classify and process data. Consequently, the structure and language design of a software development technology determines whether special data types are used and in what capacity they operate. R defines the data type as a *class*. Among the various classes used in R, a special class called a *factor* serves as both a functional capability and a data type. As a functional capability, a factor provides the technical encoding needed to distinguish quantitative from qualitative data. As a data type, it means the data has been qualitatively formalized.

FIRST PRINCIPLE:

In a world defined by data objects, a data frame is the most basic integrated data structure used in R.

Data in R is defined by the concept of an object. In the most general terms, a data object in R can represent a single data point, a list of items called a vector, a plot, or an integrated data structure referred to as a *data frame*.[4] The data frame, R's version of a spreadsheet, consists of a structured collection of information, enumerated by rows and columns. Data frames in R are either directly

---

[4]There is actually another type of data object provided in R referred to as a List object. However, a treatment of this data structure extends beyond the scope of an introductory book on R.

accessed, programmatically constructed, externally linked, or made available through data importation.

Over the years, the data frame in R has evolved by expanding in both range and functionality. Necessitated by a dynamically changing R environment, new data frame types have been developed to improve utility and applicability. Other data objects similar to data frames that are available in R include the *matrix, array, time-series objects, data tables,* and *tibbles*. A tibble is also referred to as a "simple data frame." Data tables and tibbles are modified versions of the data frame. While a treatment of matrices, time-series objects, data tables, and tibbles is beyond the scope of this book, a focus will remain on the introduction and use of the data frame.

FIRST PRINCIPLE:

Function Syntax Paradigm

Action(x , y , z) As Class

Output

arguments

Function name

R code structures are fundamentally built on the concept of a function; functions are organized into packages; a collection of packages represents a library.

There are two categories of people who use R: users and developers. Users are defined as those individuals who actively utilize the current capabilities provided by the technology. Users of R have acquired either a basic, intermediate, or an advanced knowledge of the technology, using it when needed. Conversely, developers are advanced users capable of applying an added skill set. Developers add functionality and value to the R environment in one of two ways: 1) by developing, submitting, and maintaining R packages, or 2) working directly with the R technology software itself. In an introduction to R, it is important to draw a distinction between R users and developers.

This distinction, defined by the *First Principle of the Function*, establishes a baseline for introducing a code development skill set which targets the user, not the developer.

The First Principle of the Function introduces a code concept of the function to those with little or no familiarity with the technology. While the word sounds intimidating, a function is nothing more than a pattern of words that work together to initiate an *action*. These word patterns, when deconstructed, are called *arguments*. Each argument holds a piece of data called an *input* that helps to qualify how to encode the characteristics of the action. When a function is performed it generates an *output*. Functional output in R fundamentally consists of either data objects or graphical visualizations. Finally, once a function's action has been initiated, it is said to have been "executed."

An analogy loosely based on a catapult can be used to better describe the mechanics of a function. In a visualization referred to as the *Catapult Function Analogy,*[5] the mechanics of a function are compared to a catapult's arm, its winch, and the rocks which are used as projectiles. The arm represents the function itself, serving as the launching mechanism for the rocks. The winch supports the underlying process framework through which the computation is conducted as the function is being executed. The rocks symbolize the function's arguments, which contain the inputs needed to qualify the characteristics of the function. Like the rocks in a catapult, several

---

[5]The original image can be found in the *2010 Encyclopedia Britannica*, s.v. "catapult."

arguments may rest in the "spoon" of a function. These rocks can be large or small, of a higher or lower density, or can reflect different shapes. Likewise, a function argument may require a numeric value, a qualitative measure, or special encoding unique to that argument.

Catapult Function Analogy

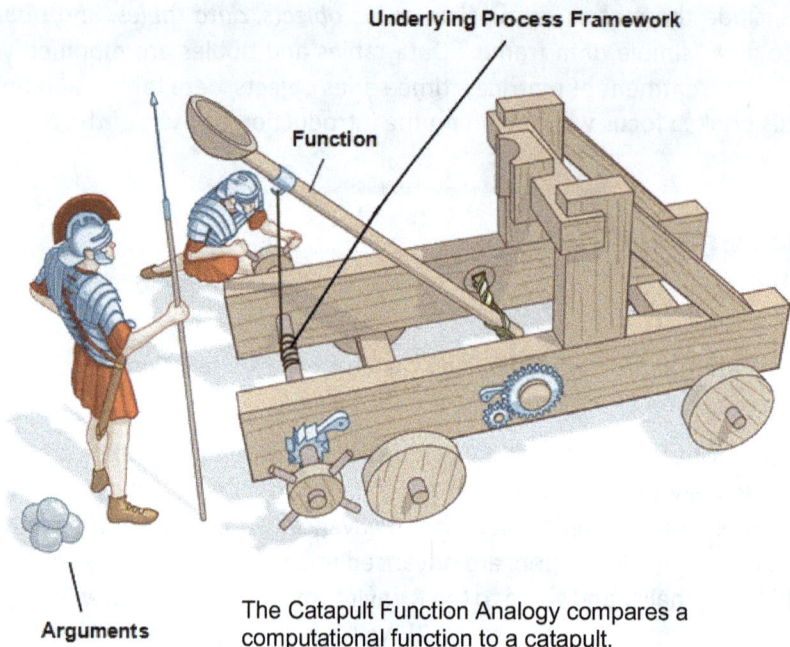

The Catapult Function Analogy compares a computational function to a catapult.

Building on the analogy of a function, let's now add another dimension for consideration. From the metaphorical to the literal, a deeper understanding of the function can be acquired through the lens of a syntax paradigm. The *Function Syntax Paradigm* shows the structural characteristics of a function, in which each part is distinguished by a different color. There are three parts of the syntax paradigm: blue defines an action, purple represents the function's arguments, and red provides an output which is of a class type that defines the data object.

Function Syntax Paradigm

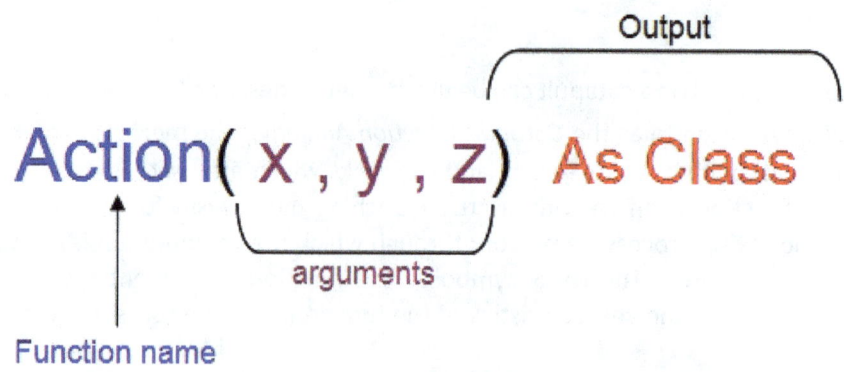

To apply the Function Syntax Paradigm to a simple example, consider a function in R called *min*. Min is short for "minima," which is a function that extracts the smallest value from a set of elements referred to as a vector. In the min function, there are two arguments to consider. The first argument requires the name of the vector of elements to be evaluated. The second argument directs the function to remove any missing values from the vector before executing the function's action.

In the example below, there are 10 values consisting of various numbers that have been converted to a numeric vector called x:

```
x = c(12, 5, 8, 13, 7, 145, 89, 100, 9, 30)
```

The "c" represents the R function required to combine multiple values together. Using the min function, we can extract the smallest number from this group. To achieve this, the following R code is applied:

```
y = min(data = x, na.rm = TRUE)
y
[1] 5
```

Transposing R's min function to the *Function Syntax Paradigm*, the following conversion with colorized annotations is provided:

```
min          =   ACTION
arguments    =   data, na.rm
values       =   x, TRUE
y            =   OUTPUT AS CLASS
```

The na.rm argument removes all missing values from the x vector, referred to as NAs in R. In this example, there were no missing values so this argument was not required. However, it was included to provide context in describing the structure of an R function.

There are additional characteristics related to the structure of an R function that should be considered. Introductory R function characteristics are listed as follows:

- The argument syntax starts with the ARGUMENT name then an = sign then the VALUE
- Commas separate each argument-value pair within a function
- Not all arguments within a function are required to generate an output
- Arguments can be defined in any order, provided that the argument name precedes the value
- Open and closed parentheses separate the arguments and their corresponding values from the function name
- R syntax is case-sensitive which means that the word *data* is different from the words *Data* or *DATA*
- If arguments are encoded in the order in which the function was designed, argument names are not required but there are exceptions
- Only the value, not the argument name is used when a function executes with a single argument

Best practices for encoding R functions is to use argument names to specifically identify the values. However, functions that are executed with a single argument implicitly reference the name of a data object. This is why the argument name is not required.

FIRST PRINCIPLE:

The primary connection between data and functions are plot visualizations.[6]

Important to an understanding of R is the role that data plays in its connection to the development of plot visualizations. While data points, vectors, and data frames provide the means by which to organize data in R, plots are used to construct a data narrative. In fact, the plot visualization capability in R lays the foundation upon which the data narrative is built. A formidable data narrative tells a compelling story. In a metaphorical play called *The Data Narrative*, its characters perform as critical insights, the story's plot evokes data sensemaking, and the climax dances to the shadows cast by the decisions informed by its own reflection.

Plot visualizations in R can be either static or interactive in form. Static plot visualizations are defined by image files that use standard graphical file extensions. The most popular graphical file extensions used in R are .bmp, .jpg, .png, and .tif. Interactive plots provide added functionality not available in a static image. An interactive plot is largely defined by the ability to use a mouse to click, hover, drag, or otherwise engage with the image in a way that produces an interactive experience between the user and the plot object. The most common file extensions used in creating interactive plots in R are .htm or .html.

The *Data-to-Plot Visualization Process Model* captures the essence of this First Principle. First, a dataset is established. Second, functional R code is written. Finally, the code is executed and a plot is created. An ensemble of plots are then designed, modeled, and organized to create an incisive data narrative. The plot provided in this example, referred to as a network visualization, reveals insights into the dataset that would have been difficult to ascertain from the data alone. This simple example demonstrates the process by which R can be used to begin translating data from a structured set of characters to a perceptive image of information.

---

[6]A secondary connection between data and functions are data models. Data models represent data that have been consolidated into organized, sub-component objects characteristic of the model. These data sub-components include data frames, vectors, and independent values. Examples of objects modeled from functions in R include Linear Regression, Histograms, and Box plots.

Data-to-Plot Visualization Process Model

Dataset → **R Code** → **Plot**

ID1	Item	CNode
1	Dale_Lesker	ProjFIN
2	Robert_Johnson	GovContract
3	Lena_Cisp	ProjFIN
4	Timothy_Sheller	Inhouse
5	Claudia_Mesker	Inhouse
6	Oscar_Fuentes	ProjFIN
7	Hadley_Bronston	Inhouse
8	Laurel_Muhammad	ProjFIN
9	Amy_Driskoll	Inhouse
10	Harold_Ramsey	Inhouse
11	Hadley_Bronston	ProjFIN
12	Robert_Johnson	ProjFIN
13	Claudia_Mesker	ProjFIN
14	Erin_Mulasky	GovContract
15	Jerry_Wheeler	ProjFIN
16	Gina_Potter	GovContract
17	Gavin_Beele	Inhouse
18	Tristan_George	GovContract
20	Jerry_Dean	OutSource
21	Noah_Price	Inhouse
22	Jim_Hamilton	OutSource

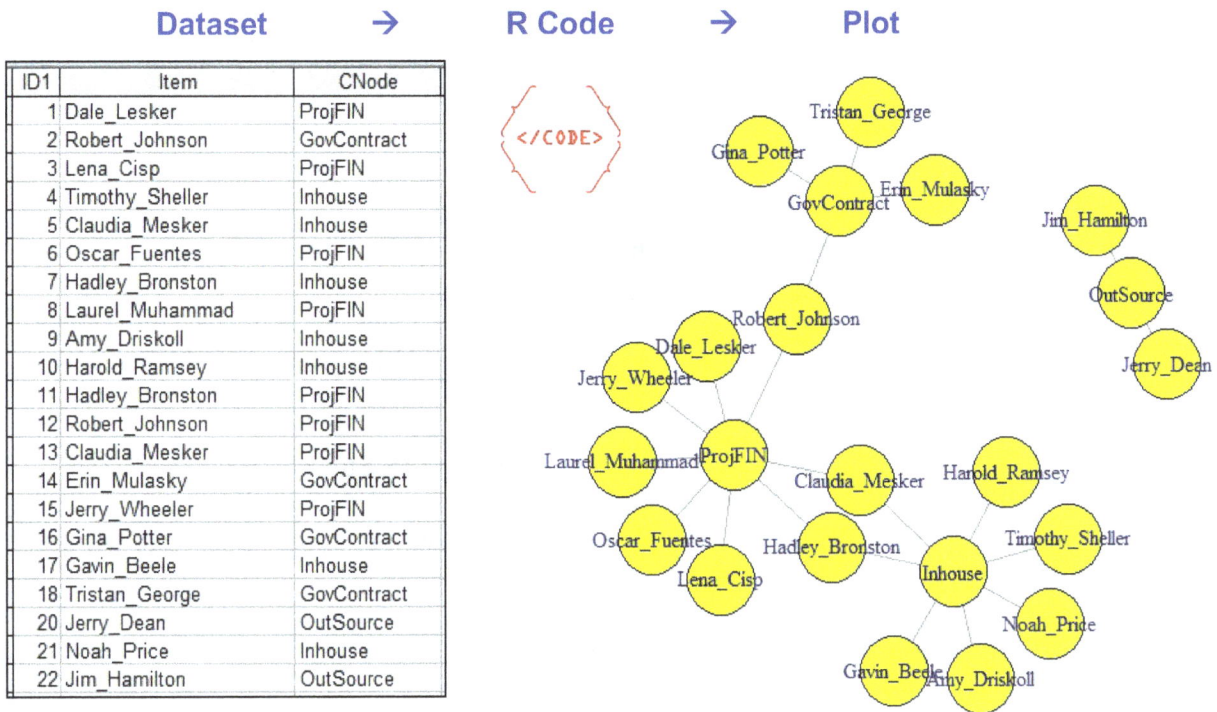

FIRST PRINCIPLE:

```
x = 1:1000000
y = x + 5
```

Quantitative data processing in R is largely based on vector mathematics.

This First Principle explores the relationship between quantitative data and its process mechanism in R. The image icon representing this First Principle contains two interrelated mathematical statements. Introduced in Chapter 1, *y* solves for *x* by abstracting the complexity from the user through a computational process called *vector mathematics*. For example, in the image icon vector *x* represents 1,000,000 independent numeric values from 1 to 1,000,000. Vector *y* computes 1,000,000 numeric values by adding 5 to each value contained in vector *x*. Vector mathematics in R applies a mathematical notation through a computational framework designed to process data. With only a few lines of R code, numerous computations can be easily encoded.

Using vector mathematics in R achieves three primary objectives:

1. Applies a convenient means of computational encoding
2. Abstracts complexity away from the user
3. Precipitates a technological gain in productivity

What if the vectors being mathematically computed are different sizes? How would these vectors be computed in R? For example, what would happen if vector *A* containing 10 numeric values was added to vector *B* which contained only 5 numeric values? See the example below:

```
A = c(6, 4, 3, 4, 3, 2, 9, 9, 4, 3)
B = c(2, 1, 3, 4, 9)
ans = A + B

ans
  [1]  8  5  6  8 12  4 10 12  8 12
```

When the equation "*ans = A + B*" is executed, the first five elements in vector B are repeated when added to the last five elements of vector A. So, the elements in vector B (2, 1, 3, 4, 9) are added to both the first five elements in vector A (6, 4, 3, 4, 3) and the last five elements in vector A (2, 9, 9, 4, 3). The results are returned in a numeric vector called *ans*, which contains a total of ten new elements.

Understanding the application of vector mathematics in R will increase the ability to correctly and decisively interpret how basic mathematical computations are encoded. Once this First Principle is understood, developing and applying vector mathematics in R will become not only possible, but inevitably successful.

NOTES

CHAPTER 3
PRACTICAL R FUNCTIONS FOR THE BEGINNER

INTRODUCTION: Chapters one and two provide both the background and context needed to acquire a general familiarity with R technology. This chapter moves to transition this familiarity from a lenified introduction to a pragmatic skill-set. Building upon the First Principles outlined in Chapter two, R technology, the code, and the methods used to interact with this technology are presented. Functional code has been strategically selected along with descriptive explanations and examples. The objective is to show you how to build a technological skill-set in R that is robust, practical, and sustainable.

PRELIMINARY SETUP: R technology, in its most popular configuration,[7] consists of two distinct but interrelated applications. These applications are referred to as the *R Console Application* and *RStudio*.[8] When used together, these applications provide an enhanced R-user experience. The R Console Application is also referred to as an advanced command-line utility. This means that the application is largely built on a framework in which the user interaction is defined by text-line commands. While the R Console Application contains a menu bar, it is essentially windowless. It is a baseline version of R. Conversely, RStudio is referred to as a shell application, which means that it sits on top of the R Console Application.

RStudio is like an accessory that proves so invaluable its emergence redefines the fashion statement. In a portrait painted with functionality, background becomes foreground.

In the same manner, RStudio's capabilities compliment the R Console Application with additional functionality, principal of which is the *Graphic-User Interface (GUI)*. A GUI-based shell allows a user to easily interact with a console application through the use of graphical objects such as windows, tabs, buttons, slider controls, scroll bars, and check boxes. While R-based activities can be conducted exclusively through the R Console Application, RStudio is easier to use and offers a better user experience. In addition, RStudio is a productivity-enhancement tool with extended capabilities beyond those provided by the R Console Application. Ultimately, the R Console Application should be used in conjunction with, not as an alternative to RStudio.

As a development tool, RStudio is referred to as an *Integrated Development Environment*, or an *IDE*. The IDE has a rich history as a software-development utility. However, as technology has advanced, the concept of the IDE has expanded its scope in new and interesting ways. RStudio, for example, is a specialized IDE generally used for statistical computing and graphics, not software engineering. According to the RStudio website:

---

[7] R technology comes in many different forms and variations. Examples include R Analytic Flow, StatET, R Tools for Visual Studio, ESS, Radiant, RBox, Revolution Analytics, and NVim-R. One of the most common configurations in R deploys RStudio, used in combination with the R Console Application.

[8] These applications are managed by two separate entities. The R Console Application is managed by the R Foundation for Statistical Computing (r-project.org) and RStudio by a company with the same name as its product (rstudio.com). RStudio is based out of the United States in Boston, Massachusetts.

"RStudio is an integrated development environment (IDE) for R. It includes a console, syntax-highlighting editor that supports direct code execution, as well as tools for plotting, history, debugging, and workspace management.

RStudio is available in open source and commercial editions and runs on the desktop (Windows, Mac, and Linux) or in a browser connected to RStudio Server or RStudio Server Pro (Debian/Ubuntu, Red Hat/CentOS, and SUSE Linux)."[9]

Consequently, R technology has evolved beyond statistical computing and graphics.

R SOFTWARE INSTALLATION: Before R technology can be used, it must be installed on either a desktop or a laptop computer. The R Console Application is installed first, followed by RStudio. Desktop versions of the software for both the R Console Application and RStudio are open source. This means the software is made available gratis, and can be downloaded directly online.

A YouTube channel hosted by a user named **Tech Decode Tutorials** provides instruction-friendly videos that detail the process for downloading and installing R software, based on the operating system you are using:

> **R for Windows Installation Video:**
> https://rebrand.ly/RstudioWindowsInstallation
>
> **R for Mac Installation Video:**
> https://rebrand.ly/RstudioMacInstallation

To view an R installation video on your phone, scan your smart phone's camera over the QR code that corresponds to the operating system running on your computer:

R for Windows

R for Mac

RSTUDIO GRAPHIC USER INTERFACE (GUI): After installing the R Console Application and RStudio, open the RStudio application. One of the most demonstrable characteristics of the application is how it is visually dimensioned. Once opened, the application should display four distinctive window panes. An open version of RStudio should look like the image captured in *RStudio Screenshot 1* (without the yellow labels).

---

[9]This content originated from the RStudio products webpage link: https://rstudio.com/products/rstudio/.

RStudio Screenshot 1

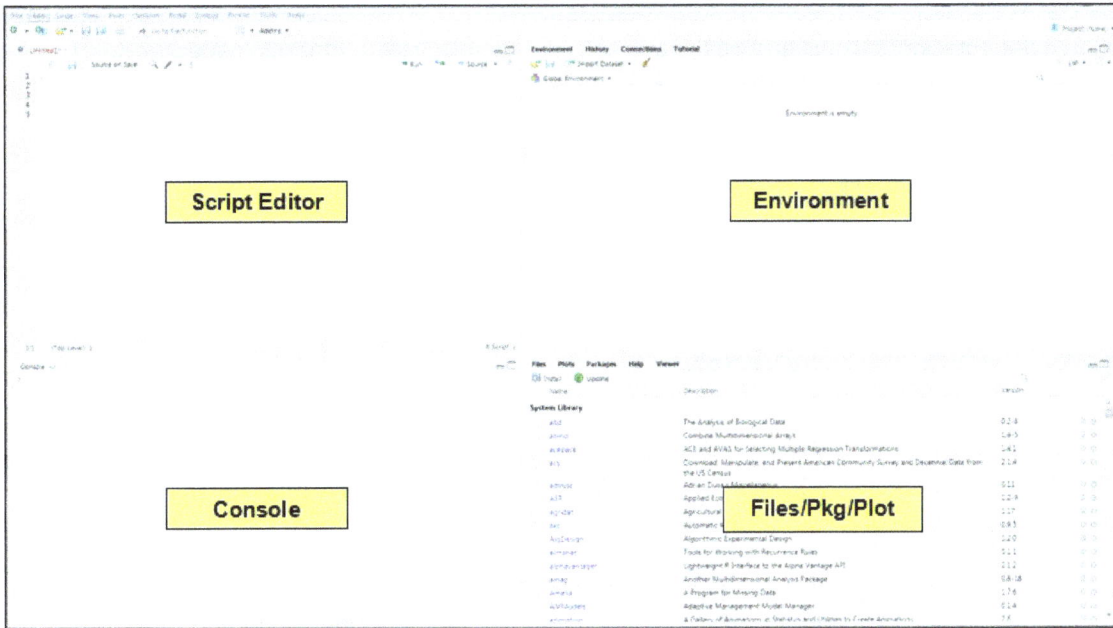

If the four window panes referenced by *RStudio Screenshot 1* are not displayed in the manner shown, the visual configuration settings will need to be changed. To modify RStudio's visual display, click the *View* menu item located at the top of the application. Ensure that the "Show All Panes" and "Console on Left" options are selected by clicking them. A blue checkmark indicates that these selections were successfully made. See the *RStudio Screenshot 2* below:

RStudio Screenshot 2

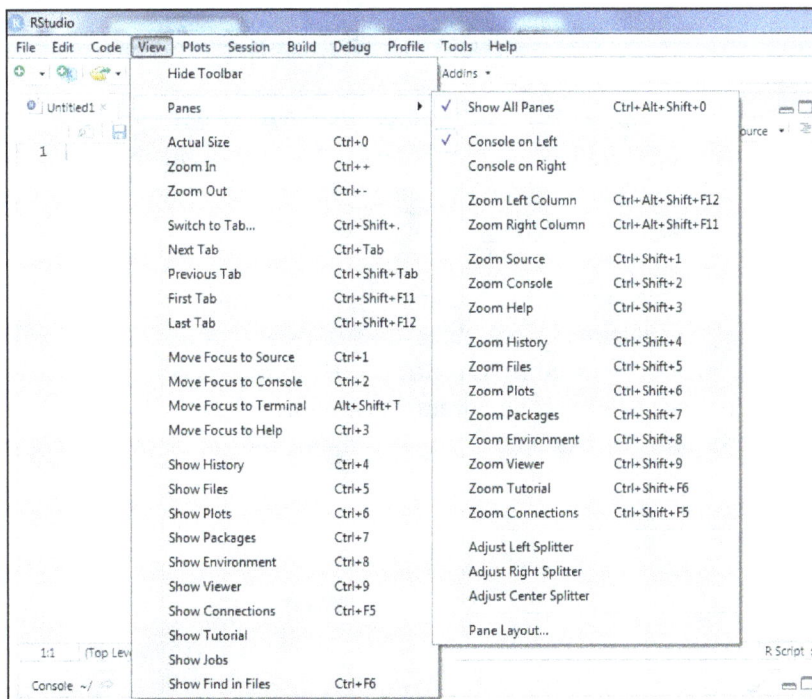

This configuration setting ensures that each of the four main window panes in RStudio are displayed and represented in the order shown in *RStudio Screenshot 1*. Visual references added to the exercises provided in this book rely on configuration settings that match this layout.

RSTUDIO GUI REFERENCE: Each of the four windows displayed in the *RStudio Screenshot 1* represent a functional component of the RStudio application. In clockwise order, each functional component is listed along with a brief description:

- **Script Editor**
  The Script Editor is a type of text tool used to create different script objects in R. For the beginning user, the Script Editor operates like a notepad on which R code, referred to as a script, is written. Notes, in the form of comments can also be added. Within this quadrant, R scripts can be developed, modeled, and stored as script files. The Script Editor is like a chalkboard for R.

- **Environment**
  This quadrant contains a listing of data objects currently available in RStudio. As data objects are created, they are stored in this quadrant. *Time spent in RStudio is called a session. A session begins when the RStudio application is opened and ends when it is closed.* Data objects can be stored between sessions. In addition, a record of all the code executed by the user can be found in a tab within this quadrant called **History**. Connections to external data and tutorial modules are also provided.

- **Files/Packages/Plots**
  Located in the lower right-hand corner of the application, this quadrant provides a set of specific capabilities. These capabilities enable plot output, file navigation, and package management. When generated, static plots appear in the **Plots** tab and html-based plots can be found in the **Viewer** tab.

- **Console**
  The final quadrant, called the Console window, is where R code is entered and subsequently executed. The word "Console" is labeled at the top of the window. In addition, this quadrant contains a blue > indicating the entry point for text. The blue > is located in the upper left-hand corner of the window.

These window quadrants, along with their underlying capabilities, define the essence of RStudio. To assist in developing a familiarity with RStudio, a visual reference has been added to supplement the exercises. A 2x2 matrix, referred to as a ***GUI Matrix***, provides a visual cross-reference when working between RStudio windows. Quadrants shaded in yellow call attention to the RStudio window being referenced. For example, the RStudio *GUI Matrix* below directly references the Console window:

RStudio GUI Matrix

BEGINNER R FUNCTIONS: This section provides a listing of (40) critical beginner functions used in R. Each function applies a code sample represented within a structured exercise. Each exercise builds on the knowledge introduced by previous exercises. These functions are not only commonly used, but are inherently flexible in utility across a wide range of tasks and projects. To execute the R code provided in these exercises, follow these general steps:

- Type the code provided in the exercise
- Press the Enter key on the keyboard after each code line

As these exercises are being practiced, the Console window will begin to fill up with code lines and output results. To clear the Console window, either click the gray brush icon located at the top of the Console window, or press the Ctrl+L keys on the keyboard.

GUI Matrix

1. Function: c

The first step in building a formidable skill-set in R is to learn how to create a basic data object. The following code stores a single data item into a data object called *dobj*:

```
dobj = "United States"
```

GUI Matrix

The data object called *dobj* should now be listed in the Environment window. To view the contents of the *dobj* object, type "dobj" in the Console window then press the Enter key.

```
dobj
```

[1] "United States"

GUI Matrix

To add more than one item to a data object, referred to as an element, the c function is used:

```
dobjs = c("United States", "Canada", "Mexico")
```

Combining multiple elements in R creates a data structure referred to as a *vector*. The c function can be used to combine any number of data elements together. It can also be used to combine multiple data elements within a functional argument. Numeric values can be included in a data object's name but cannot be its first character. For example, a data object called *Day2* is acceptable but *2Day* is not. Data objects are also case-sensitive. This means that data objects with the names *Day2*, *day2*, and *DAY2* represent three distinct objects in R.

2. Function: rm

This function permanently removes one or more data objects from the Environment window.

```
rm(dobj)
```

GUI Matrix

3. Function: data.frame

A data frame is the most basic integrated data structure used in R. Its scope extends beyond a single data element or a vector. Defined as R's version of a spreadsheet, a data frame consists of a structured collection of information, organized by columns and rows. Creating a data frame object in R requires three steps:

1) Establish column names
2) Populate the data frame
3) Call the *data.frame* function

The following code is used to create a basic data frame in R:

```
df = data.frame(total = 1:5, fruit = c("orange", "grape", "banana", "pear", "apple"))
```

To see the code results, press the Enter key. Now go to the Environment window and click the spreadsheet icon located on the right side of the *df* object. The data frame output is now displayed in the Script Editor window as a separate tab called *df*.

GUI Matrix

The data frame function contains two arguments – one for each field name. In this example, the argument is the field name itself. The *total* field name applies the syntax "1:5." The colon can be substituted for the word "through." So, this code syntax can be interpreted as "one through five."

Another way to create a data frame is to separately define a set of vectors, after which they are programmatically connected through the *data.frame* function:

```
patientID = 1:4

age = c(25, 34, 28, 52)

status = c("Poor", "Improved", "Excellent", "Poor")

patient_data = data.frame(patientID, age, status)
```

Notice the difference in how data frame and vector objects are organized within the Environment window. Data frames and model objects are located in the **Data** section. Vectors and single data item objects can be found in the **Values** section. This distinction is important as it helps to visually differentiate between data object types in RStudio.

GUI Matrix

Since the values have already been defined, they are not directly included in the data.frame function.

When manually typing code in the Console window, only one line of code can be executed at a time. To execute multiple lines of code simultaneously, there are two options:

OPTION 1:
- Type the code on a tab called *Untitled1** in the Script Editor window
- Copy (Ctrl+C) and paste (Ctrl+V) the code to the Console window
- Press the Enter key on the keyboard

GUI Matrix

OPTION 2:
- Type the code on a tab called *Untitled1** in the Script Editor window
- Place the cursor at the end of the first line of code
- Click the Source button located at the top of the *Untitled1** tab

After executing the code from the Console window, go to the Environment window and click the spreadsheet icon located on the right side of the *patient_data* object. The data frame output is now displayed in the Script Editor window as a separate tab called *patient_data*.

GUI Matrix

HELPFUL TIPS:
The following information provides additional insights into the usability of RStudio:

- To clear the Console window, click the gray brush or press the Ctrl+L keys on the keyboard.
- To access code previously entered into RStudio, click the blinking cursor in the Console window. Then press the ↑ Arrow key on the keyboard. Each press of this key scrolls through the history of code previously executed. Accessing code in this manner prevents having to re-type commands.

GUI Matrix

- To review a list of all code commands that have been executed since the last purge, click the **History** tab found in the Environment window.
- When executing multiple lines of code in R, **OPTION 1** is the preferable option to use. **OPTION 1** details a process which records the executed lines of code in the **History** tab. **OPTION 2** executes the lines of code within a function called *source*, so the code lines per se are not directly accessible in the **History** tab.
- To change the order of the tabs listed in the Script Editor, click on the tab to be reordered and drag it laterally to a desired position. To close a tab, click the "x" located after the tab name.

GUI Matrix

- This exercise consists of multiple code lines. It is the first step in building a skill set used to develop R scripts.

4. Function: fix

There are two ways to modify data within a data frame. Data can be modified by either using RStudio's Data Editor, or programmatically by applying R code. The *fix* function opens the Data Editor obviating the need to change specific data points programmatically. Column names can also be modified using the Data Editor.

In this exercise, the Data Editor will be used to modify data points contained in the *df* data frame created in Exercise 3. The R code below programmatically opens the Data Editor:

GUI Matrix

```
fix(df)
```

Once the Data Editor opens, select the cell to be modified by clicking it. Once selected, enter the new data. Change the 2 in the total column to 12 then change the word "apple" to "mango" in the fruit column. Once the changes have been made, close the Data Editor window. The *df* data frame object has now been updated.

Before Changes

Data Editor			
File Edit Help			
	total	fruit	var3
1	1	orange	
2	2	grape	
3	3	banana	
4	4	pear	
5	5	apple	
6			
7			

After Changes

	total	fruit
1	1	orange
2	12	grape
3	3	banana
4	4	pear
5	5	mango

To programmatically change the data points in a data frame requires its rows and columns to be directly referenced. Code is used to first reference the row number then the column number. For example, using the code syntax below, the numeric value of 5 provided in the *total* column will be changed to 8.

GUI Matrix

```
df[5,1] = 8
```

Consider the code syntax as a programmatic rule that says, "change the value of row number 5, column 1 to a value of 8."

5. Function: library

The R System Library is a collective repository of frameworks representing a stored set of functional capabilities. These frameworks are more formally referred to as R packages. A package is a critical component of R's System Library. In fact, the System Library exemplifies much of the functionality defined by R technology.

A collection of these packages can be found in an online repository called the *Comprehensive R Archive Network*, or *CRAN*.[10] CRAN is hosted by the *R Foundation for Statistical Computing*. CRAN contains thousands of R packages that can be downloaded and used in RStudio.

However, before the functional capabilities of an R package can be used, it must be downloaded, installed, and loaded from RStudio's System Library. It is important to recognize that each of these actions are distinctive. Definitions of these actions help to explain the nuances in the relationship between the accessibility and usability of an R package. A brief definition of these actions are provided as follows:

---

[10]See https://cran.r-project.org/web/packages/available_packages_by_date.html.

- *Package Download:* The electronic transfer of an R package file, typically through the use of an internet connection, from a website to a computer.
- *Package Installation:* The successful configuration of a new R package in an R-compatible software application.
- *Package Load:* The successful activation of an installed R package.

GUI Matrix

An initial installation of RStudio includes a System Library with a pre-installed set of R packages.[11] A listing of these packages can be found in the Files/Pkg/Plot window under the **Packages** tab.

A function in R cannot be actively used if the package supporting it has not been loaded. However, a package cannot be loaded before it has been installed. Unless global or starting configurations have been modified, seven R packages are loaded each time an RStudio session is opened. The functions referenced in this chapter generally use R packages that have been both installed and pre-loaded. Specific guidance is provided when packages are used beyond those initially loaded at the beginning of each user session.

INSTALL A NEW R PACKAGE:
To install a new R package, click the *Install* button on the Packages tab in the Files/Pkg/Plot window.

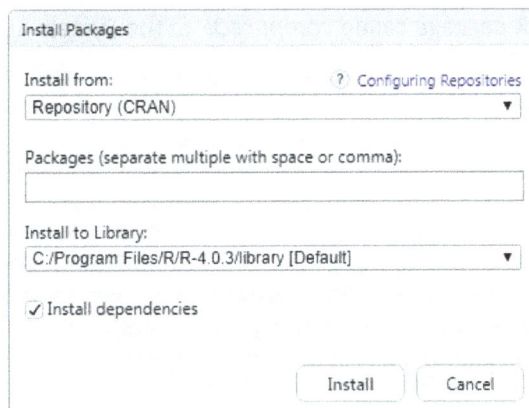

```
Install Packages

Install from:                    ? Configuring Repositories
Repository (CRAN)                              ▼

Packages (separate multiple with space or comma):
[                                              ]

Install to Library:
C:/Program Files/R/R-4.0.3/library [Default]   ▼

✓ Install dependencies

                        [ Install ]  [ Cancel ]
```

Keep all the recommended defaults and type the name of the R package in the *Packages* text field. As an example, type the R package **DescTools**. Remember, R package names are case-sensitive. After typing the package name, click the *Install* button located at the bottom of the dialog box. The package should successfully install. Unless a package is corrupt or has been previously removed, installing a package is required only once.

LOAD AN R PACKAGE:
After installing the **DescTools** package, it must be loaded before it can be used. To load an R package, use the following code:

```
library(DescTools)
```

---

[11]Only a small subset of R packages available on CRAN are installed.

To verify that the **DescTools** package has been both successfully installed and loaded, go to the System Library located under the **Packages** tab. Using the scroll bar, scroll down the list to the **DescTools** package entry. **DescTools** should be located in the list along with a black check mark next to the package name.

GUI Matrix

A listed package name means the package has been successfully installed. A package name complimented with a black check mark means the package has been loaded and is ready to use. *A package must be loaded once per session before it is used.*

As an alternative to code, manually managing package loading is made available in RStudio. Manually selecting or deselecting a check box in RStudio's System Library controls package loading without having to write R code.

REMOVE AN R PACKAGE:

To permanently remove an R package from the System Library, click the button located on the same line as the R package to be removed. Then confirm the action.

GUI Matrix

PROGRAMMATICALLY INSTALL & REMOVE AN R PACKAGE:

To programmatically install or remove an R package, use the *install* and *remove* functions. The code below installs an R package called *compareDF* to the System Library then removes it.

```
install.packages("compareDF")

remove.packages("compareDF")
```

UPDATE THE SYSTEM LIBRARY:

R packages are consistently being updated. New functions in packages are added, old functions are removed, and existing functions are refined or extended. It is important to have an updated System Library. With moderate R use, the System Library should be updated on a weekly basis.

GUI Matrix

To update RStudio's System Library, go to the **Packages** tab located in the Files/Pkg/Plot window. Click the green button called *Update* and the dialog box shown below will appear. Click the *Select All* button located at the bottom of the window, then click *Install Updates*. The window will close and then R code will be displayed as it is executed in the Console window. After the code runs, package updates will be complete and the System Library will be current.

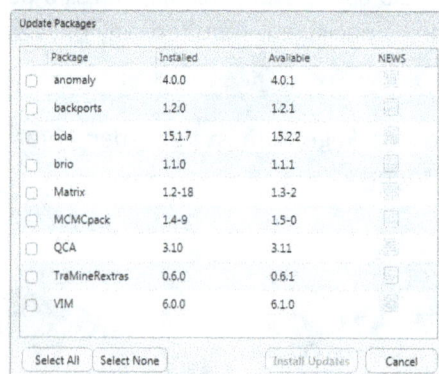

Update Packages			
Package	Installed	Available	NEWS
anomaly	4.0.0	4.0.1	
backports	1.2.0	1.2.1	
bda	15.1.7	15.2.2	
brio	1.1.0	1.1.1	
Matrix	1.2-18	1.3-2	
MCMCpack	1.4-9	1.5-0	
QCA	3.10	3.11	
TraMineRextras	0.6.0	0.6.1	
VIM	6.0.0	6.1.0	

Select All   Select None   Install Updates   Cancel

6. Function: data

A data analysis cannot be conducted without data. In Exercise 3, the *data.frame* function was introduced, demonstrating how to create a dataset. In addition to customized datasets, R offers a variety of pre-defined datasets. These datasets can be used to sharpen the coding skills needed to effectively analyze and plot data.

Pre-defined datasets in R are accessed programmatically by using the *data* function. The *data* function extracts a user-selected dataset from a target package directly into the RStudio application. R provides a special package called *datasets* that contains 60 sample datasets, all of which are accessible and easy to use. In addition to these datasets, thousands of other datasets are available within various R packages provided on CRAN.

To access a dataset from the datasets package, type the word "data" then an open parenthetic symbol (Shift+9) in the Console window. Once entered, a series of dataset options will appear in a pop-up menu list along with the package name where they can be found. This feature is referred to as *intellisense*. It is a code-completion capability designed primarily to increase productivity by facilitating accurate code writing. However, intellisense also provides supplemental information about the code being used. Using the scroll bar, scroll through the dataset listing to determine which dataset to select. Once identified, click the dataset name then press the Enter key on the keyboard. The dataset will now appear as a data object in the Environment window. If the newly created data object contains a label that reads, "*<Promise>*," then it needs to be categorically formalized. To formalize the new data object, type the first few characters of the dataset name in the Console window. The target dataset is now available to use.

GUI Matrix

Intellisense in RStudio

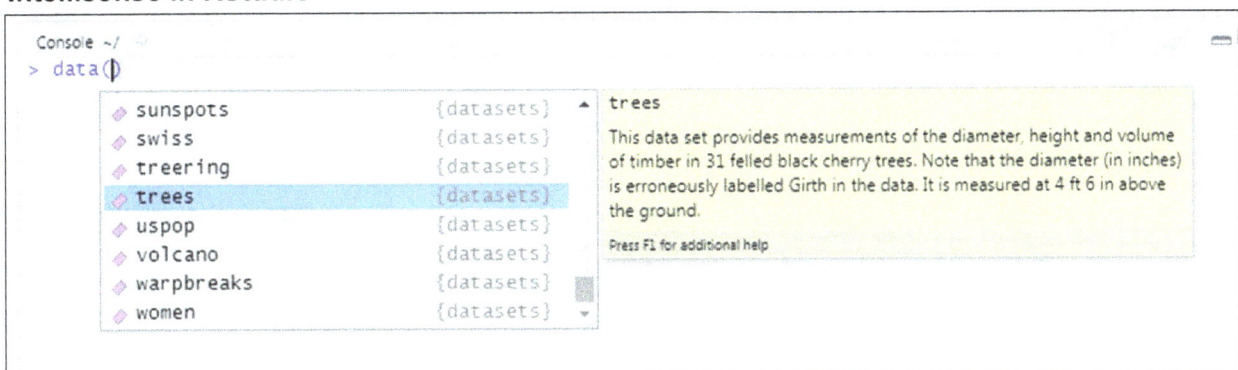

```
Console ~/
> data()
```

sunspots	{datasets}	trees
swiss	{datasets}	
treering	{datasets}	This data set provides measurements of the diameter, height and volume of timber in 31 felled black cherry trees. Note that the diameter (in inches) is erroneously labelled Girth in the data. It is measured at 4 ft 6 in above the ground.
trees	{datasets}	
uspop	{datasets}	
volcano	{datasets}	Press F1 for additional help
warpbreaks	{datasets}	
women	{datasets}	

To get more information about a dataset follow these steps:

- In the pop-up menu list, hover over the dataset of interest; a yellow title tip providing additional information about the dataset will appear.
- While hovering over a dataset of interest, press the F1 key on the keyboard. Then go to the **Help** tab found in the Files/Pkg/Plot window for additional details about the dataset.

GUI Matrix

```
data("trees")

data("Orange")

data("PlantGrowth")
```

7. Function: names

The *names* function retrieves or modifies the column names of a data frame object. Building on the *data* function introduced in Exercise 6, the column names from a dataset called *attitude* will be extracted into a vector object called *fldnames*.

```
data("attitude")

fldnames = names(attitude)
```

GUI Matrix

Two new data objects called *attitude* and *fldnames* should now be added to the Environment window. To see the results of the *fldnames* object, type the word *fldnames* in the Console window then press Enter on the keyboard.

```
fldnames
```

[1] "rating"   "complaints" "privileges" "learning" "raises"   "critical" "advance"

Since each record in the *attitude* dataset represents the average scores taken from a questionnaire of 35 employees, the *rating* column name should be changed. The following code changes the column name from *rating* to *avg rating*:

```
names(attitude)[1] = "avg rating"
```

The *rating* column name has now been changed to *avg rating*. The number one in brackets represents the column number of the column name to change. The next action is to update the fldnames vector by executing that line of code a second time.

```
fldnames = names(attitude)

fldnames
```

[1] "avg rating" "complaints" "privileges" "learning" "raises"   "critical" "advance"

Practice using the *names* function by changing the column names of each field in the *attitude* dataset.

8. Function: cbind

Short for "column-bind," the *cbind* function combines similarly structured vectors or columns of data into one data object. It is an effective function used in reconstructing or repurposing data. Using the *cbind* function requires row counts to be the same between the vectors being combined. For example, a vector of 50 items cannot be combined with one containing 25 items. Data objects to be combined must contain the same row count.

In this exercise, columnated data taken from two separate datasets will be combined into a new dataset called *newds*. The two datasets used are *BOD* and *Formaldehyde*. The *Time* data column from the BOD dataset will be combined with the *carb* data column in the

Formaldehyde dataset. To achieve this objective, a reference to a column name must be made within the code. When both a data object name and a columnated reference are called together within a function's argument, the $ symbol is used as a separator. As a separator, the $ activates an *intellisense* pop-up menu list that displays all the column names that define that object.

```
data("BOD")
data("Formaldehyde")
newds = cbind(BOD$Time, Formaldehyde$carb)
newds = data.frame(newds)
```

GUI Matrix

Remember that R is case sensitive. To access R's *intellisense*, type the first few characters of BOD in capitalized letters. A drop-down menu list will appear. Scroll and then hover over the BOD option and either click the mouse or press the Tab key on the keyboard. Then, type the $ and another menu list will appear that displays the BOD dataset's column names. Click the Time column. Repeat this action for the Formaldehyde dataset then press the Enter key to execute the code. The final line of code converts the *newds* data object into a data frame.

GUI Matrix

Once the *newds* dataset has been created, go to the Environment window and click the spreadsheet icon located on the right side of the *newds* object. The output is now displayed in the Script Editor window as a separate tab called *newds*.

Notice that due to the conversion action, the column names have changed. Use the code provided in Exercise 7 to change the field names back to *Time* and *carb* respectively.

9. Function: sample, set.seed, and sort

The *sample* function provides a mechanism for building customized datasets. As a result, it is a powerful data simulation builder. The *sample* function can generate vector data based on specific characteristics including uniqueness, size, and probability. Used by beginners and advanced users alike, the *sample* function generates pseudo-randomized data of all types. Simulated data can be constructed and shaped using binary, numeric, categorical, or alphanumeric characters. Considering the scope of its output, the economy of code applied by this function is remarkable.

This is the first exercise in the series in which a function's arguments are formalized with specificity. In this exercise, four examples are provided that showcase the encoding flexibility of the *sample* function:

Example 1:
In this example, the sample function is used to generate a simulated categorical data vector containing 25 items:

```
vec1 = sample(x = c("shirt", "shoes", "pants", "socks", "jacket"), size = 25, replace = TRUE)
```

Example 2:

In Example 2, the sample function is used to generate a numeric distribution of 25 simulated values between 1 and 100, none of which are duplicated in the output:

```
vec2 = sample(x = 1:100, size = 25, replace = FALSE)
```

Example 3:

In Example 3, the sample function is used to generate a binary distribution of 25 simulated values with a distribution ratio of 20% zeros and 80% ones respectively:

```
vec3 = sample(x = 0:1, size = 25, replace = TRUE, prob = c(.2, .8))
```

The arguments generally used in the *sample* function are described as follows:

x	=	One or more elements that define the data vector
size	=	The total element size of the data vector
replace	=	A boolean value indicating whether the items defined by argument x can be duplicated
prob	=	The probability distribution for each categorical group defined by argument x

Example 4:

The final example combines all three of the previous examples into a single data frame object. The *set.seed* function is added to generate data reproducibility, preserving both the form and order of data. This function uses a numeric argument called *seed*. The seed value represents the function's starting point for generating pseudo-random data. The *kind* argument establishes the type of Pseudo Random Number Generator (PRNG) to use.

```
set.seed(seed = 21, kind = "Mersenne-Twister")

vec1 = sample(x = c("shirt", "shoes", "pants", "socks", "jacket"), size = 25, replace = TRUE)

vec2 = sample(x = 1:100, size = 25, replace = FALSE)

vec3 = sample(x = 0:1, size = 25, replace = TRUE, prob = c(.2, .8))

simds = data.frame(vec1, vec2, vec3)
```

GUI Matrix

Using the *head* function, the first six records from the *simds* data frame object are displayed:

```
head(simds)
    vec1   vec2   vec3
1   shirt   73    1
2   pants   42    0
3   shirt   53    0
4   shoes   92    1
5   jacket  80    0
6   pants    5    1
```

A review of the *vec2* data shows a set of numeric values. One way to begin understanding the distribution of these values is to sort them. The *sort* function can be used to organize both numeric and non-numeric data in either ascending or descending order.

sort(simds$vec2)

[1] 5 7 8 9 37 40 42 44 45 50 53 56 58 63 65 66 68 69 72 73 74 80 81 91 92

To sort a data field in descending order, add an argument to the *sort* function called "decreasing." This argument calls for a boolean value of either True or False. The default order is ascending, so the argument is not included in the aforementioned code. However, a descending sort requires the argument to be included and set to True.

GUI Matrix

sort(x = simds$vec2, decreasing = TRUE)

[1] 92 91 81 80 74 73 72 69 68 66 65 63 58 56 53 50 45 44 42 40 37 9 8 7 5

Alternatively, data can be quickly and flexibly sorted in RStudio without using code. Go to the Environment window and click the spreadsheet icon located on the right side of the *simds* object. The data frame output for *simds* is now displayed in the Script Editor window in a tab with the same name.

GUI Matrix

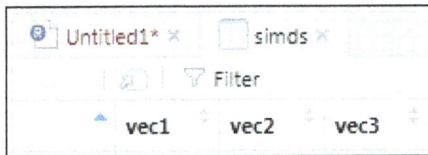

Each data field has its own sorting button. Click either △ or △ to sort by the field of choice.

HELPFUL TIPS:
The following information provides additional insights into the usability of the *sample* function:

- When encoding the *sample* function, arguments *x*, *size*, and *replace* should always be used. The *prob* argument is optional.
- When used, the *prob* argument generates approximated probability distributions.
- To reproduce output provided by the *sample* function, use a function called *set.seed*. It should be executed *before* the sample function.
- To get more information about an RStudio function, type the ? symbol then the function name. For example, additional information about the sample function can be found by typing *?sample* in the Console window, then pressing the Enter key. Once executed, go to the **Help** tab found in the Files/Pkg/Plot window.

GUI Matrix

10. Metadata Functions: dim, length, ls, class, str, summary, and Mode

This section explores a series of functions designed to extract data about data. "Metadata," to which the term is commonly referred, describes or defines a unique characteristic about data. Collectively, these characteristics shape metadata discovery. Each metadata function provides a technical dimension to the value proposition of data. The value proposition of metadata is measured by its context. For example, the size of a dataset impacts the quality of a data analysis. A data analysis supported by a dataset consisting of 50 columns and 500,000 records will be very different from a data analysis conducted from an abridged version of the same dataset.[12] Metadata can also be used as a means to expand a contextual understanding of

---

[12] An example of an abridged version of the dataset might consist of 25 columns and 10,000 records.

data. Context can be extended to explain the size justification of a dataset, the conditions under which the data were collected, and the process used to manage and verify data accuracy.

In another example of metadata context, the impact relative to the number and types of data objects used in a data project is considered. The extent to which data is selected, processed, and modeled impacts a project in subtle but distinctive ways. As data objects are defined, workflow design and process modeling are impacted. These activities invariably impact the project scope which then directly impacts the project budget. Irrespective of the causal order of impact, these interconnected project relationships exist. Understanding metadata context provides another lens through which the value proposition of data can be assessed.

Metadata Functions Used in R

This exercise uses the same datasets and vector objects created in previous exercises. Enter the code lines in the Console window then press the Enter key on the keyboard.

- dim: Shorthand for *dimension*, this function returns the size dimensions of a data frame object in row and column terms. In R, rows are referred to as *observations* and columns as *variables*. The dimensions for each data object can also be found in the middle of each line item under the Data section in the Environment window.

GUI Matrix

```
dim(simds)
```

[1] 25 3

- length: Used to get the total number of items contained in a vector. When used with a data frame, the total number of columns are returned.

```
length(fldnames)
```

[1] 7

GUI Matrix

```
length(simds)
```

[1] 3

- ls: Shorthand for *list object*, this function returns a listing, in alphabetical order, of all the data objects contained in the Environment window. In the example provided below, the *ls* function returns all the data object names created in this chapter from Exercises 1 through 9.

GUI Matrix

```
ls()
```

[1] "age" "attitude" "BOD" "df" "dobjs" "fldnames"

[7] "Formaldehyde" "newds" "Orange" "patient_data" "patientID" "PlantGrowth"

[13] "simds" "status" "trees" "vec1" "vec2" "vec3"

Notice there are no arguments used in the *ls* function. While arguments exist for this function, they are not used in this example.

R Functions can be nested. This means that functions can be executed within functions to generate output. A simple example nests the *ls* function inside the *length* function. The output returns the total number of list objects generated by the *ls* function:

length(ls())

[1] 18

- class: This function returns the dimension type of a data object based on its encoding.

  Data object dimensionality in R typically classifies data into two categories. One category dimensionalizes data by structure, and the other category, if applicable, by data type. Some data objects, however, are unidimensional by construction. This means that the data object can only be dimensionalized by its data type not by a structure as defined by R. For example, a vector is a collection of elements and a dataset is a collection of vectors. A vector is considered to be a unidimensional data object in R. Conversely, a dataset, comprising a series of vectors each of which represents a different dimension of itself, serves as a multi-dimensional data object. To programmatically show the dimensional nuances in data, the following code returns the structural dimension of the *simds* data object:

  class(simds)

  [1] "data.frame"

  GUI Matrix

  However, if the class function is encoded by vector, it returns a different result:

  class(simds$vec1)

  [1] "character"

  In the second example, a dimension called *vec1* is defined as a character data type. The *vec1* dimension is a component part of a larger multi-dimensional data object called *simds*.

  Data object dimensionality plays a critical role in R. Knowing the difference between a data structure and a data type improves the ability to successfully encode functions, data models, and plots in R.

- str: The *str* function, short for structure, builds on the concept of data object dimensionality described by the *class* function. This function returns a listing of both the structure and data types comprising a data frame object. For example, when applied to the *simds* data object, the *str* function returns the following results:

```
str(simds)
```

```
'data.frame':    25 obs. of  3 variables:
$ vec1: chr  "shirt" "pants" "shirt" "shoes" ...
$ vec2: int  73 42 53 92 80 5 72 37 91 9 ...
$ vec3: int  1 0 0 1 0 1 0 1 1 1 ...
```

GUI Matrix

str Output Explanation:

- ♦ THE FIRST LINE
  - ➢ Returns the structural dimension of the simds data object, including the number of records (observations) and columns (variables) comprising the data frame

- ♦ SECOND AND SUBSEQUENT LINES
  - ➢ Starts with a dimension separator ($)
  - ➢ Next is the column name (vec1, vec2, and vec3) followed by a colon
  - ➢ The next component represents the data type; in the example provided, the data types are either "character" or "integer;" the most common data types include chr, Date, Factor, int, logi, and num
  - ➢ The last component returns a sample of the data represented by its data type

Alternative Means of Access

In RStudio, there is an alternative means by which to access data object dimensionality without code. Using RStudio's Graphic User

GUI Matrix

Interface (GUI), clicking the ▷ button displays information about a data object similar to that returned by the *str* function.

Click this button located next to the *simds* line item in the Environment window.

After the button has been clicked, it changes its appearance to a ▽ button, unveiling data type information as it unfolds within the window. Clicking the button again folds the data up.

- summary: The final function in the metadata group is *summary*. This function generates what is called a *descriptive statistic* of the data object. A descriptive statistic is a summary statistic that quantitatively describes or summarizes features from a collection of information.[13] The *summary* function applies a *descriptive statistic* to get a quick overview of a data object in its aggregated form. It is R's version of a Pivot Chart. The code below applies the *summary* function to extract a descriptive statistic from the *simds* dataset:

---

[13]See https://en.wikipedia.org/wiki/Descriptive_statistics.

summary(simds)

```
     vec1             vec2              vec3
 Length:25       Min.   : 5.00    Min.    :0.00
 Class :character 1st Qu.:42.00   1st Qu. :1.00
 Mode  :character Median :58.00   Median  :1.00
                  Mean   :53.92   Mean    :0.76
                  3rd Qu.:72.00   3rd Qu. :1.00
                  Max.   :92.00   Max.    :1.00
```

GUI Matrix

Applying a descriptive statistic to a numerically defined vector returns six critical metrics about the data. Examples of numeric vectors are represented in both *vec2* and *vec3*. These metrics describe the vector in terms of its quartiles, numeric range, and what are referred to as general *Measures of Centrality*. Quartiles organize the vector into four groups in ascending numeric value. The numeric range identifies the smallest and largest values in the vector; *Measures of Centrality* capture the mean and median values.

The example below more specifically describes the descriptive statistic of numeric values captured by the *summary* function:

Min.	=	The smallest value in the vector
1st Quartile	=	The first interval of data also referred to as the 25th percentile
Median	=	The second interval of data also referred to as the 2nd quartile or the 50th percentile
Mean	=	The average value in the vector
3rd Quartile	=	The third interval of data also referred to as the 75th percentile
Max.	=	The largest value in the vector

There is one more *Measure of Centrality* called the *Mode* that is not included in the summary function output. The *Mode* returns both the most frequent value and its frequency of occurrence in a vector. A function called *Mode* found in the DescTools package can be used to capture this information.

library(DescTools)

Mode(simds$vec1)

[1] "shoes"

attr(,"freq")

[1] 7

GUI Matrix

Mode(simds$vec2)

[1] NA

attr(,"freq")

[1] 1

```
Mode(simds$vec3)
```

[1] 1

attr(,"freq")

[1] 19

"Shoes" is the most frequent value found in the *vec1* data with a frequency occurrence of 7. Since the *vec2* data was initially created with unique values, the Mode for that vector is 1. "NA" in R is considered to be a missing value indicator or alternatively designated as "Not Available." Finally, the numeric value of 1 is the most frequent element found in *vec3* with a frequency occurrence of 19.

11. Data Type Functions: as.character, as.factor, as.logical, as.integer, as.numeric, levels, nlevels, and unclass

The exercises in this section build on the metadata functions provided in Exercise 10. These examples introduce the ability to encode datum based on its attribute. These attributes are grouped within a set of programmatic capabilities called *Data Type functions*. For example, the *simds* dataset, introduced in Exercise 9, contains three columnated data fields, or vectors. Each vector is encoded as a class, alternatively referred to as a *data type*. These vectors are encoded based on an attribute and the context in which they are used. To that end, data encoding in R is a type of technical formalization.

Before vectors are modeled or plotted, they must be correctly encoded. Incorrectly encoded vectors in R can produce inaccurate or unanticipated results. In the case of chart plotting, referencing the wrong vector or data type in a function argument may return errors that prohibit the chart from being generated. Whether vectors are imported or created, the data type defining each vector must be evaluated and verified.

Given this requirement, important questions emerge. How do you evaluate a vector to determine its data type? What are the rules used to encode data? How do you know if a vector has been correctly encoded?

Correctly evaluating data to be considered for encoding begins by applying the First Principle 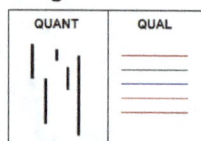. Data in R are generally represented as either a quantitative or a qualitative measure. This information can be used to determine how data should be encoded.

In R, numeric values that represent a measure, total, or count are commonly associated with quantitative data. Quantitative data are typically encoded using the *as.numeric* function. Conversely, qualitative data are categorical in nature. Data that can be categorized should be encoded using the *as.factor* function. While there are exceptions, these recommendations provide a general baseline to follow. From the results provided by the *str* function and the *simds* summary from Exercise 10, let's evaluate the data types to see if they are correctly encoded.

```
     vec1              vec2              vec3
 Length:25         Min.   : 5.00     Min.   :0.00
 Class :character  1st Qu.:42.00     1st Qu.:1.00
 Mode  :character  Median :58.00     Median :1.00
                   Mean   :53.92     Mean   :0.76
                   3rd Qu.:72.00     3rd Qu.:1.00
                   Max.   :92.00     Max.   :1.00
```

By applying the *str* function used in Exercise 10, we can see that vec1 is encoded as class character. Both vec2 and vec3 are encoded as class *int*. Numeric data in R are commonly encoded as either a *num* or an *int* class data type. *Int* is an abbreviation for *integer*. Integers are encoded by using the *as.integer* function. When creating quantitative data, it should be encoded using the *as.numeric* function. However, if the data have already been generated, for example through importation, then a conversion from class *int* to class *num* is unnecessary. The difference between the two data types involves how the data are stored in the computer's memory. In addition to *as.integer* and *as.numeric*, there are other numeric data types used in R. However, this exercise will apply the most common function used in R to encode numeric data – *as.numeric*.

Data Type Evaluation: vec1
The vec1 data object comprises a series of clothing articles that are encoded as class character. Is vec1 correctly encoded? It is clear that this data is not quantitative but qualitative in nature. It contains no numeric values. It is not unreasonable to believe that vec1 is correctly encoded. However, before a final determination can be made, another question must be asked. Can the data elements in vec1 be categorically defined? In this case, articles of clothing including shirts, shoes, and pants represent potential data categories. Is there a relationship between these data categories and any other data vector provided by the *simds* dataset? Let us assume that the numeric values in vec2 describe quantities of clothing articles in vec1. This assumption would confirm that vec1 is categorical and should therefore be encoded as class factor and not as class character. To re-encode vec1 as a factor, apply the *as.factor* function:

simds$vec1 = as.factor(simds$vec1)

GUI Matrix

To confirm that vec1 has been correctly encoded, verify its class type:

class(simds$vec1)

[1] "factor"

Data Type Evaluation: vec3
While numeric like vec2, vec3 incorporates an additional characteristic. The data it holds is binary in form. This means that vec3 data is represented as either 1s and 0s, or by the terms "True" or "False." As a result, binary data in R can be class-encoded in one of several ways. It can be encoded as a factor, a character, a numeric, or as a logical class. In what capacity binary data is encoded largely depends on how the data is being used. For example, if vec3 was going to be used in a machine learning project, it would most likely be encoded as class numeric. If the *simds* dataset were going to be used in a plot, vec3 might be encoded as a factor. Finally, if vec3 required specific tagging as a True or False value, it would be encoded as class logical.

In this exercise, vec3 will be encoded as a factor. Using the *as.factor* function, vec3 is re-encoded from class numeric to class factor:

```
simds$vec3 = as.factor(simds$vec3)
```

To illustrate the relationship between binary data and variously encoded data types, a new column of data called vec4 will be appended to the *simds* dataset. It will replicate the data in vec3 but will be encoded differently.

GUI Matrix

To append a new column to a data frame, enter the name of the data frame to which the column will be appended followed by $. Then type the new column name. In this example, the new column name is "vec4."

```
simds$vec4 = vec3
```

The vec4 column replicates both the data and the original data type in vec3 which is class integer. To clarify, there are two vec3 data objects related to this exercise. One version of vec3 is contained within the *simds* dataset and is of class factor. Additionally, there is the vec3 data object that represents a separately defined vector of class integer. Both of these vec3 data objects can be found in the Environment window.

GUI Matrix

The data type in vec4 will now be changed to class logical. The re-encoded change in the data type will also change the data content from 0/1 to TRUE/FALSE values.

GUI Matrix

```
simds$vec4 = as.logical(simds$vec4)
```

To review the first six records of the modified *simds* dataset, apply the head function:

```
head(simds)
   vec1 vec2 vec3  vec4
1  shirt   73    1  TRUE
2  pants   42    0 FALSE
3  shirt   53    0 FALSE
4  shoes   92    1  TRUE
5 jacket   80    0 FALSE
6  pants    5    1  TRUE
```

GUI Matrix

Notice that the 1s in vec3 have been converted in vec4 to "True" values and the 0s have been converted to a corresponding value of "False." When converting binary data to class logical, the binary data type to be converted should be either of class integer or class numeric. Converting binary data from class factor to class logical will generate NAs, returning an unexpected and inaccurate result.

Exercise 11 executed structural and metadata changes to the *simds* dataset. Let's now compare the dataset's descriptive statistics before and after data type changes were made. For example, extracting a descriptive statistic from the *simds* dataset in Exercise 10 returns the following results:

Exercise 10 *simds* Summary Results

```
    vec1              vec2            vec3
Length:25         Min.    : 5.00   Min.    :0.00
Class :character  1st Qu.:42.00    1st Qu.:1.00
Mode  :character  Median :58.00    Median :1.00
                  Mean    :53.92   Mean    :0.76
                  3rd Qu.:72.00    3rd Qu.:1.00
                  Max.    :92.00   Max.    :1.00
```

After conducting an evaluation of the *simds* dataset, the data types of vec1 and vec3 were re-encoded to correctly align the data with its data type. The *simds* dataset was structurally changed by adding a new data vector called vec4. By executing the *summary* function against *simds*, the following results (with a gratuitous caption added) are returned:

summary(simds)

Exercise 11 *simds* Summary Results   **GUI Matrix**

```
     vec1           vec2          vec3        vec4
jacket:3      Min.    : 5.00    0: 6    Mode :logical
pants :6      1st Qu.:42.00     1:19    FALSE:6
shirt :6      Median :58.00             TRUE :19
shoes :7      Mean    :53.92
socks :3      3rd Qu.:72.00
              Max.    :92.00
```

The differences in the encoded results are clearly stated and precisely defined.

Extracting Factor Characteristics

Three key functions are used to extract data characteristics specific to a factor-encoded vector. These functions are *levels*, *nlevels*, and *unclass*.

- levels: This function returns an alphabetical listing of the categorical names from a data field encoded as a factor data type:

  levels(simds$vec1)

  [1] "jacket" "pants" "shirt" "shoes" "socks"

- nlevels: Similar to the levels function, nlevels returns the total number of categories comprising a data field encoded as a factor data type:

  GUI Matrix

  nlevels(simds$vec1)

  [1] 5

- unclass: The unclass function converts categorical data into a corresponding set of numeric categories. The enumeration is determined by the alphabetic order of the categories. In addition, it extracts the encoded labels from the vector's factor data type:

  unclass(simds$vec1)

  [1] 3 2 3 4 1 2 2 5 4 2 4 2 5 1 4 3 4 4 1 5 4 2 3 3 3
  attr(,"levels")
  [1] "jacket" "pants" "shirt" "shoes" "socks"

GUI Matrix

To better visualize how categories are numerically converted, another vector called "vec5" will be added to the *simds* dataset:

  simds$vec5 = unclass(simds$vec1)

A review of the first six records comprising the modified *simds* dataset shows the following:

  head(simds)

```
    vec1 vec2 vec3  vec4 vec5
1   shirt   73    1  TRUE    3
2   pants   42    0 FALSE    2
3   shirt   53    0 FALSE    3
4   shoes   92    1  TRUE    4
5  jacket   80    0 FALSE    1
6   pants    5    1  TRUE    2
```

12. Measures of Centrality Functions: mean and median

In Exercise 10, the concept of the *Measure of Centrality* was introduced. One way to identify *Measures of Centrality* within a dataset is to apply the summary function. However, to independently extract these measures from a targeted vector of numeric values, the mean and median functions are used:

mean(simds$vec2)

[1] 53.92

For example, the mean, or average of the values represented in the vec2 vector is 53.92. Conversely, the median, or the 50th percentile of the same data vector returns 58:

GUI Matrix

median(simds$vec2)

[1] 58

13. General Measures of Variability Functions: var, sd, range, min, and max

To capture the extent to which numeric values are dispersed within a vector, three functions are generally used. The function var measures the variance, or the average of the squared differences measured from the mean. The standard deviation, represented as the sd function in R, measures the distance a set of numeric values are spread out from the mean. Finally, the range function returns both the smallest and largest values in a numeric vector. Using the vec2 vector, each of these *Measures of Variability* are computed:

var(simds$vec2)

[1] 652.8267

sd(simds$vec2)

[1] 25.55047

range(simds$vec2)

[1] 5   92

To independently extract either the smallest or the largest numeric value from a vector, the min and max functions are used:

min(simds$vec2)

[1] 5

max(simds$vec2)

[1] 92

14. Summation and Product Functions: sum and prod
The ability to effectively add and multiply a series of numbers is a fundamental skill set in R. The function used for addition is the sum function:

sum(c(45, 2, 10))

[1] 57

To multiply a series of numbers together, the prod function is applied:

prod(c(10, 3, 4, 7, 9))

[1] 7560

Arithmetic operations in R apply the same symbols typically used in software engineering, mathematics, finance, and statistics:

+ = Addition
- = Subtraction
* = Multiplication
/ = Division

10+5   10-5   10*5   10/5
[1] 15   [1] 5   [1] 50   [1] 2

15. Measures of Association: cor

A correlation, also referred to as a *Measure of Association*, is a term used in Statistics to describe a statistically significant relationship between two dependent variables. While there are many factors to consider in evaluating, testing, and verifying correlations, the process generally begins by using the cor function. This function establishes a general baseline for identifying a correlation.

In this exercise, correlations will be computed for the girth, height, and volume from a dataset called "trees." The code below extracts a copy of the trees dataset then computes the correlations for each variable pair.

```
data("trees")

cor(x = trees$Girth, y = trees$Height)
```

[1] 0.5192801

```
cor(x = trees$Girth, y = trees$Volume)
```

[1] 0.9671194

```
cor(x = trees$Height, y = trees$Volume)
```

[1] 0.5982497

GUI Matrix

The *x* and *y* arguments in the cor function represent the numeric vectors to be assessed. Generally, the larger the correlation, the more positively correlated the variables are likely to be. In the example provided, the relationship between tree girth and volume returns the most positively correlated results among the variables measured at around 97%.

To generate correlations without specifying a correlation function for each variable pair, a *Correlation Matrix* can be used. A *Correlation Matrix* is designed to organize the results of all pairwise correlation combinations within a dataset. To successfully generate a *Correlation Matrix*, all values in the dataset must be numeric.

```
cor(trees)
           Girth     Height    Volume
Girth  1.0000000 0.5192801 0.9671194
Height 0.5192801 1.0000000 0.5982497
Volume 0.9671194 0.5982497 1.0000000
```

GUI Matrix

16. Applying Simple Conditional Expressions: filter and ifelse

Being able to extract subsets of data from a dataset facilitates flexibility and provides a means for partitioning a dataset. Creating new data fields by stratifying data adds value to a dataset. Programmatically validating data ensures its integrity. The commonality shared by each of these tasks is a *Simple Conditional Expression.* A *Simple Conditional Expression* applies logic operators to a set of rules, called expressions. These expressions are then evaluated against a data vector to return either a true or a false statement. The "simple" part of the term derives its name from the binary nature of the results. A principal component of the *Simple Conditional Expression* is the logical operator which defines the relationship between the expression and the result. R provides a set of common logic operators that can be used to construct a *Simple*

Conditional Expression. These logic operators are outlined in the **R Logic Operator Chart**. This exercise provides three classic examples of *Simple Conditional Expressions*.

R Logic Operator Chart

LOGIC OPERATOR	DESCRIPTION
<	Less Than
>	Greater Than
<=	Less Than or Equal to
>=	Greater Than or Equal to
==	Equal to
!=	Not Equal to
!	Not
&	And
|	Or

- **filter**: The base R package includes a function called *subset* that can be used to partition data. However, the *filter* function found in the dplyr package is preferred, and much easier to use. Using the filter function, multiple conditions can be applied, extending its partitioning capability.

- **ifelse**: This function uses a basic If-Then logic structure to return a binary result between one of two alternatives. Alternative outputs are not limited to true or false literal values.

In this example, the following conditions are encoded in the *filter* function to extract a subset of data from the *simds* dataset:

- Data in the vec1 column must contain the word shirt
- The numeric value in the vec2 column must be greater than 50
- The logical value in the vec4 column must be TRUE

The results returned *must meet each* of these conditions.

install.packages("dplyr")

library(dplyr)

filter(.data = simds, vec1 == "shirt" & vec2 > 50 & vec4 == "TRUE")

```
  vec1 vec2 vec3 vec4 vec5
1 shirt   73    1 TRUE    3
2 shirt   56    1 TRUE    3
3 shirt   65    1 TRUE    3
```

GUI Matrix

To store the results in a variable of class data frame, provide an object name followed by an equal sign at the beginning of the filter line. Notice that the second argument in the *filter* function **is** the *Simple Conditional Expression*. Consequently, the *filter* function is unorthodox in its expression as the second argument does not require an argument name.

In a second example, a *Simple Conditional Expression* is used in a dataset called *trees* to categorize a tree as either "small"[14] or "tall." The results are then appended to the dataset as a new data field called *Cat*. A numeric value of 76 baselines the category. What is the significance of 76? This value represents the average tree height, serving as the baseline distinguishing between a small or large tree.

---

[14]The term "short" could have been used but "small" is a better descriptor. Context matters.

```
mean(trees$Height)
```

[1] 76

```
trees$Cat = ifelse(test = trees$Height > 76, yes = "TALL", no = "SMALL")
head(trees)
```

GUI Matrix

```
  Girth Height Volume   Cat
1   8.3     70   10.3 SMALL
2   8.6     65   10.3 SMALL
3   8.8     63   10.2 SMALL
4  10.5     72   16.4 SMALL
5  10.7     81   18.8  TALL
6  10.8     83   19.7  TALL
```

Adding the *Cat* field stratifies the trees dataset. If one understands the data contained in a dataset, data stratification can be used as an effective means to expand its context. In conjunction with this expansion, increasing the value proposition of a dataset is not only possible, but probable.

In the final example, data validation captures the highest scores recorded for either a tester's IQ or the results of a language test taken by the tester. The objective is to identify high scores in both categories. The dataset used in this exercise contains three fields of data: lang, IQ, and method. Using a dataset called *teach.df* taken from an R package called s20x, a *Simple Conditional Expression* can be applied to meet the challenge posed by the objective.

```
install.packages("s20x")
```

GUI Matrix

```
library(s20x)
```

```
data("teach.df")
```

```
teach.df$h_lang_IQ = ifelse(test = teach.df$lang >= 90 | teach.df$IQ >= 135, yes = "TRUE",  no
= "FALSE")
```

Results are provided in an appended data field called *h_lang_IQ*. In this exercise, the rules defining the *Simple Conditional Expression* are as follows:

- A value in the *lang* column of the teach.df dataset is greater than or equal to 90

    OR...

- A value in the *IQ* column of the teach.df dataset is greater than or equal to 135

If either of these conditions is true then the result is true. Otherwise the result returns a false value. Notice the vertical line that separates the two test conditions. This character is referred to as a "pipe." It is used in R as the conjunction for OR.

To evaluate the results in context, the teach.df dataset will be returned in a specific sort order.

In Exercise 9 Example 4 of this chapter the sort function was introduced. It executes a sort against a single data vector. However, to return an entire dataset in which one or more data columns have been sorted requires using a different function. The Sort function found in the lessR package provides the sorting capability needed for this task. Since R is case sensitive, the sort function, as it is represented in both its lower and higher case forms, reflects two separate functions. Each function is applied differently, containing its own set of arguments. The code below applies the higher case Sort function, which adds context to the result.

GUI Matrix

```
install.packages("lessR")

library(lessR)

head(Sort(data = teach.df, by = "IQ", direction = "-"))
```

```
Sort Specification
  IQ -->  descending

   lang  IQ method h_lang_IQ
9    91 139      1      TRUE
28   70 138      3      TRUE
7    92 135      1      TRUE
6    86 133      1     FALSE
12   98 128      2      TRUE
16   90 125      2      TRUE
```

The nested code extracts the first six records of the modified teach.df dataset, sorted by IQ in descending order. There are three arguments used in the Sort function: *data*, *by*, and *direction*. The *data* argument defines the dataset used. The *by* argument references the name or names of the columns to sort. The *direction* argument applies either a + or - symbol to control the sort method. Sorting can be conducted in either an ascending (+) or a descending (-) order.

In each of the first six records of the dataset, with the exception of one record, values in either the lang or the IQ field met the conditions established by the *Simple Conditional Expression*. In two cases shown, both the lang and IQ values matched the conditional expression (rows 9 and 7). However, row number 6 did not meet either of these conditions. This is indicated by the FALSE value in the h_lang_IQ field. The lang value was not equal to or greater than 90 nor was the IQ of the tester equal to or greater than 135.

To preserve the sort order, store the results in a variable.

17. Missing Values: NA
Working with, and managing missing values in a dataset is an important part of data management. Not all datasets are data complete. As you work with data, inevitably you will encounter a dataset with missing information. The question is, what is an effective strategy for managing missing data in a dataset? R provides several packages and functions that can be used to address this challenge. This exercise will introduce the most fundamental strategies and options available for managing missing data.

A constant called "NA" is considered to be the standard missing value indicator used in R. As an acronym, NA can be interpreted as "Not Available." Missing values correctly formatted in an R dataset appear as an italicized grey *NA*.

Representations of Missing Data

Missing data within a dataset can be represented in many forms. The most common ways in which unknown or missing data are represented in a typical dataset are as follows:

- As some form of literal representation of NA including "NA" or N/A
- As a 0
- As an empty value
- Some other uniquely encoded missing or unknown value representation

Before re-encoding missing data in R, missing values should be confirmed with the data originator. Every attempt should be made to convert missing data to known values. Knowing how missing values are represented plays a critical role in being able to correctly format an otherwise incomplete dataset. For example, zeros do not necessarily represent missing or unknown values. Ultimately, if missing data cannot be reconciled with known values, working with an incomplete dataset will be unavoidable.

There are fundamentally two steps that should be followed when working with an incomplete dataset. *First, identify then verify all forms of missing data represented in the dataset. Second, properly normalize the dataset by encoding all missing values as italicized gray NAs.* In this exercise, a particularly egregious set of missing and inaccurately encoded values are included in a simulated dataset called survey. The objective is to identify and convert all missing value representations to the standard NA missing value constant.

This exercise applies coding functions and structures introduced in earlier exercises. The following code creates a simulated survey dataset in which 20 responses have been recorded for three questions. Each survey response represents a numeric value between 0 and 4:

```
set.seed(seed = 272, kind = "Mersenne-Twister")

Q1 = sample(x = 0:4, size = 20, replace = TRUE)

Q2 = sample(x = 0:4, size = 20, replace = TRUE)

Q3 = sample(x = 0:4, size = 20, replace = TRUE)

survey = data.frame(Q1, Q2, Q3)
```

GUI Matrix

Now, let's add random but reproducible missing values to the survey dataset. The following code adds three missing N/A values to the Q1 column and five empty values to the Q3 column:

```
set.seed(seed = 1367, kind = "Mersenne-Twister")

survey[sample(x = 1:20, size = 3, replace = FALSE), 1] = "N/A"

survey[sample(x = 1:20, size = 5, replace = FALSE), 3] = ""
```

The survey dataset is now considered to be an incomplete and poorly formatted dataset. Under no circumstance should a dataset configured in this fashion be used to conduct any kind of data analysis. Irrespective of how a dataset with these characteristics is generated, a significant format problem exists. In fact, in its current state, the survey dataset's format is so grossly compromised that not even the most basic computations can be performed against it.

For example, an attempt to compute the mean of the survey dataset's Q1 column returns a warning message prohibiting the computation:

GUI Matrix

mean(survey$Q1)

[1] NA

Warning message:
In mean.default(survey$Q1) :
    argument is not numeric or logical: returning NA

Executing a summary calculation in the Q3 column containing an assortment of numeric and empty values produces a similar error:

sum(survey$Q3)

Error in sum(survey$Q3) : invalid 'type' (character) of argument

Additional research on the survey dataset has revealed that there are only one of four valid responses for each question. A 0 response indicates that the question was not answered. In this example, 0 represents another form of missing data. A review of the survey dataset shows that both the Q2 and Q3 columns contain missing data in the form of a 0. Missing data is also represented in Q1 by "N/A" values and in Q3 by empty values.

To resolve these formatting inconsistencies, the survey dataset must be normalized. This means correctly encoding all missing values, irrespective of form, with NAs. The code below applies a *Simple Conditional Expression*, introduced in Exercise 16, to achieve this task:

survey$Q1 = ifelse(test = survey$Q1 == "N/A", yes = NA, no = survey$Q1)

survey$Q2 = ifelse(test = survey$Q2 == 0, yes = NA, no = survey$Q2)

GUI Matrix

survey$Q3 = ifelse(test = survey$Q3 == "", yes = NA, no = survey$Q3)

survey$Q3 = ifelse(test = survey$Q3 == 0, yes = NA, no = survey$Q3)

These *Simple Conditional Expressions* can generally be interpreted as follows:

> *"If a data point in the referenced data vector is equal to a value specifically defined as a missing value, then replace that value with an NA constant. Otherwise retain the original value."*

Each representation of a missing value in the survey dataset encoded as a N/A, 0, or an empty value have all been successfully encoded as NAs. The last step in successfully finalizing the survey dataset's format is to convert the Q1 and Q3 data types from class character to class numeric. No data type conversion is required for Q2 as its original missing value was 0. This value is consistent in its format with a data type of class integer. By applying the structure function, the data types comprising the survey dataset show the following:

str(survey)

GUI Matrix

```
'data.frame':   20 obs. of  3 variables:
 $ Q1: chr  "4" "3" NA "2" ...
 $ Q2: int  1 1 3 3 3 2 4 1 4 1 ...
 $ Q3: chr  NA "3" "4" "2" ...
```

The following code converts the Q1 and Q3 data types from class character to class numeric:

```
survey$Q1 = as.numeric(survey$Q1)

survey$Q3 = as.numeric(survey$Q3)
```

Applying the structure function a second time verifies the conversion results:

```
str(survey)
'data.frame':    20 obs. of  3 variables:
 $ Q1: num  4 3 NA 2 1 4 2 1 4 1 ...
 $ Q2: int  1 1 3 3 3 2 4 1 4 1 ...
 $ Q3: num  NA 3 4 2 2 NA NA 3 4 3 ...
```

GUI Matrix

The survey dataset has now been successfully normalized as an incomplete dataset.

Executing Basic Computations Against Vectors with Missing Data

An important distinction needs to be made between a complete and an incomplete dataset when basic computations are being performed. Basic computational functions in R are configured differently between complete and incomplete datasets. Basic computations performed against an incomplete dataset require an additional functional argument called *na.rm*. This argument informs the function that missing values exist, and must be removed before the computation can be successfully executed. The additional argument is unnecessary when a basic computation is applied against a dataset with no missing values.

The code sample below shows how a function for a basic computation is configured between vectors with and without missing values. The first step is to create the vectors.

Let's create a vector with no missing values (nmv):

```
nmv = c(10, 2, 45, 78, 93)
```

GUI Matrix

The following vector contains two missing values (mv):

```
mv = c(10, NA, 45, 78, NA)
```

The code below returns the basic computation for the mean measured against the nmv vector object:

```
mean(nmv)
```

[1] 45.6

The same function is now applied to the mv vector, with the *na.rm* argument added:

```
mean(x = mv, na.rm = TRUE)
```

[1] 44.33333

The *na.rm* argument is added to the mean function to correctly compute vectors containing missing values. The argument is a shortened form of the phrase "NA remove." If the *na.rm* argument is omitted from a basic computation in which it is required, the function will return an incomplete result:

mean(mv)

[1] NA

The list below provides a general collection of R functions that include the *na.rm* argument. It should be noted that this argument is exclusive to general-purpose functions used in R. There is a wide range of variability in the structural design of functions across the R platform. Missing data may be handled implicitly or explicitly, depending on function design. In some cases, missing data may be prohibited from being used. Before using a function, it is always a good practice to review its argument list and any syntax examples provided.

Basic Computational R Functions that Use the na.rm Argument

R Functions Using na.rm
cor
max
mean
median
min
Mode
prod
range
sd
sum
var

Conclusion
Achieving a high degree of competence in R means developing a skill set consistent with the following themes:

- Practice using functions on simulated data to see the results
- Move to develop a programmatic skill set influenced by familiarity and repetition
- Reach a comfortable familiarity with both the context and the capabilities in which R functions can perform
- Acquire an applied understanding of code syntax construction
- Know how to apply a wide range of functions in different packages across the R platform

<u>NOTES</u>

CHAPTER 4
UNLOCKING THE DATA NARRATIVE

What does it mean to *unlock a data narrative*? It is a turnkey phrase that describes the conversion of data from a dataset to a graphic visualization. R offers a plethora of ways in which data can be examined, processed, and stored. However, one of its premier capabilities centers around graphic visualization. In unlocking the data narrative, R technology provides an extraordinary capacity for converting data into what are commonly referred to as plots. These plots can be generated as high-quality digital images or as web-based graphics called *Hypertext Markup Language (html)* files.

The plotting possibilities in R are so extensive that an appropriate treatment on the topic would require its own book.[15]

The classic package used to develop plots in R has historically been *ggplot2*. However, this package maintains a steep learning curve and is designed to support advanced plotting capabilities. There should be a way in which a synoptic examination of plot mechanics can be taught that meets the needs of the beginner. Creating a plot for beginners learning R should be practical and easy to apply but also visually effective.

Plot development in R has evolved over the years making it easier today to generate plots than in the past. The reason is that several new packages have been introduced in R that facilitate plot development without using *ggplot2*. Consequently, the trend in plot development has been for package developers to integrate independent plotting capabilities within their packages. Package examples containing independent plotting features include akc, DataExplorer, DecisionAnalysis, DescTools, trajectories, and TraMineR. Alternatively, packages have been created with a focus on developing unique and interesting plots with functional ease. Package examples in this category include ggpubr,[16] ezplot, flametree, shipunov, and WVPlots. Across the R platform, packages are replete with functions designed to generate a variety of compelling plots supporting data analyses of all types.

This chapter explores plot development in R from a beginner's perspective. It is designed to defer the complexities commonly associated with advanced plot mechanics. Before pursuing advanced plotting capabilities using packages like *ggplot2*, an examination of fundamental plot syntax and code structures must first be understood. In addition, the context of use is provided to explain the relationship between data, the plot object, and the functional structure of applied code.

This chapter examines seven of the most common plot types used in R to include the following:

- Bar Plot
- Box Plot

---

[15]One of the most popular books devoted to the subject of graphical visualizations in R was authored by Winston Chang. The book is called *R Graphics Cookbook (2013)*. It is approximately 383 pages in length.

[16]Although this package applies the *ggplot2* package in its plot mechanics, the implementation is transparent to the user.

- Fan Plot
- Histogram
- Line Plot
- Mosaic Plot
- Scatter Plot

Simulating A Dataset

To illustrate the relationship between a dataset and a plot object, a dataset in R must first be either created or referenced. In this exercise, a simulated dataset called *wid* will be created. It describes the performance metrics of a simulated widget. The dataset is a data frame containing 50 records and 5 data columns: *measure, runs, tester, type,* and *mod*.

To begin, create the simulated *wid* dataset by applying the following code:

```
set.seed(seed = 145, kind = "Mersenne-Twister")

measure = abs(rnorm(n = 50))

runs = sample(x = 1:7, size = 50, replace = TRUE)

tester = as.factor(sample(x = c("Burke", "Fenton", "Logan", "Osland", "Wilks"), size = 50, replace = TRUE, prob = c(.4, .1, .1, .2, .2)))

type = as.factor(sample(x = c("Airflow", "Heat Var", "Particulate"), size = 50, replace = TRUE))

mod = as.factor(sample(x = c(5340, 2089, 6880, 1254), size = 50, replace = TRUE))

wid = data.frame(measure, runs, tester, type, mod)
```

In creating the *wid* dataset, two new functions are introduced:

- *abs* – Converts a number to its absolute value. This means that all numeric values passed through this function return a positive number. Negative values are converted to positive values. For example, applying *abs* to -0.0352 returns 0.0352. Positive values remain the same.
- *rnorm* – This function generates a normal distribution of 50 numeric values. These values, defined by the argument *n*, are computed by an approximation of the mean and standard deviation of the distribution. Both operations represent arguments in the function. In the example, the *rnorm* function used to create the measure vector applies the default values of 0 for the mean and 1 for the standard deviation, implicitly defined.

A brief description of each vector comprising the *wid* dataset is as follows:

- **measure** – A numeric value used as the average performance measure of the widget. The values are created by simulating a normal distribution of data.
- **runs** – The number of times the widget was performance-tested. The total number of runs for each tested widget ranges from 1 to 7.
- **tester** – The last name of the tester who conducted the widget performance test. Five testers are included in the *wid* dataset. By applying the prob argument, probabilities were added to control the frequency with which testers were distributed across the vector. The frequency distribution on the low end is 10%; on the high end it is 40%.

- **type** – Describes what the widget measures. The widgets in the *wid* dataset measure either airflow, heat variance, or particulate matter.
- **mod** – Defines the widget's model number. Model numbers provided in the *wid* dataset are 1254, 2089, 5340, and 6880.

To review the first six records of the *wid* dataset, use the head function:

```
head(wid)
```

```
    measure runs tester      type  mod
1 0.6869129    2 Fenton   Airflow 5340
2 1.0663631    2  Wilks  Heat Var 5340
3 0.5367006    2  Burke  Heat Var 2089
4 1.9060287    2  Wilks  Heat Var 5340
5 1.0631596    4 Osland   Airflow 5340
6 1.3703436    1 Fenton   Airflow 2089
```

Notice that three of the five vectors were encoded as class factor including *tester, type,* and *mod*. Each of these vectors share a common characteristic: they are categorically defined. Testers can be grouped by name, the type can be classified by measure, and mod can be categorized by its model. This characteristic indicates that each vector is qualitative in form. Qualitative data is generally encoded in R as class factor.

Understanding the Data Plot Relationship

Before pursuing a deeper examination of plots in R, it is important to understand how data is structured on a plot surface. Basic plots display fundamentally two dimensions of data, also referred to as vectors in R. When these data dimensions are plotted, a symbiotic relationship is created. The symbiosis between these two data dimensions is manifested on the plot's x and y axes. The x axis data dimension is plotted laterally. Relative to the x axis, the y axis dimension is plotted vertically. For example, suppose there are five categories of data plotted laterally on the x axis. Each category represents a vertical rectangle. On the y axis are a set of numeric values vertically positioned. In this example, the height of each rectangle plotted on the x axis would be determined by its corresponding value represented on the y axis.

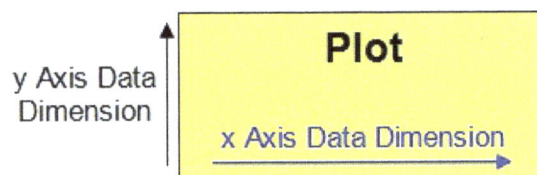

Bar Plot

A Bar Plot is a graphic visualization that uses rectangular bars to measure, by a proportion of its values, a set of categorical data. The rectangular bars can be presented either vertically or horizontally. Using the *wid* dataset, two different Bar Plot types can be generated. One can be created from an *itemization* of data in which subgroups are counted and categorized. Another approach summarizes data by using a concept called *aggregation*. Aggregation extracts data into subgroups based on an arithmetic operation conducted on the numeric values of another vector. This exercise presents an example for each of these Bar Plot types.

To create a Bar Plot by *Itemization*:

```
plot(x = wid$type, xlab = "Type", ylab = "Count", col = "green4", main = "Widget Measurement Type Bar Plot (n=50)")
```

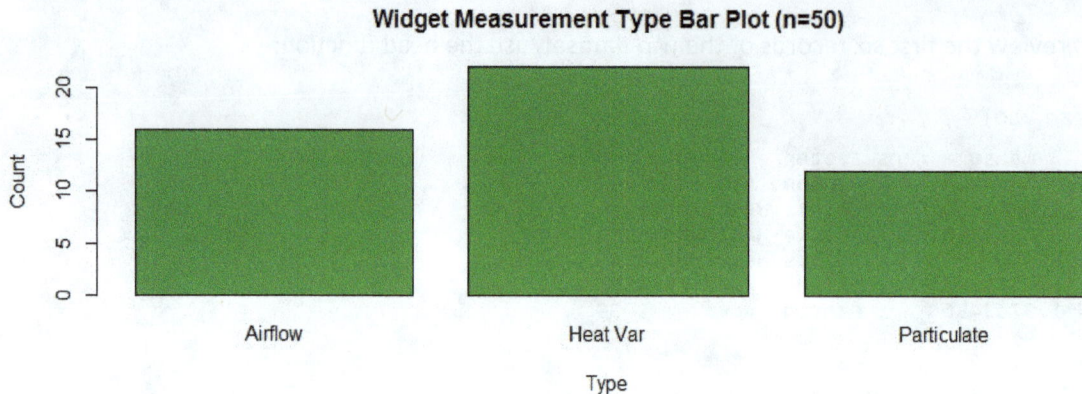

This Bar Plot is generated with one line of code. It shows the total counts from an itemized listing of widget measurement types extracted from the *wid* dataset. To get the exact counts for each measurement type, apply the summary function:

```
summary(wid$type)
```

Airflow	Heat Var	Particulate
16	22	12

The *plot* function is used to generate the Bar Plot. The code syntax for this plot applies five key arguments. Each of these arguments are briefly explained:

- **x** – The vector reference used to generate the plot. The vector must be of class factor.
- **xlab** – The label name of the plot's x axis.
- **ylab** – The label name of the plot's y axis.
- **col** – The color(s) used for the rectangular bars.
- **main** – Sets the Bar Plot's title. The total number of samples comprising the dataset being plotted should also be included in the title. This is encoded as "n=[total sample number]."

To create a Bar Plot by *Aggregation*:

```
ag = aggregate(formula = runs ~ type, data = wid, FUN = "sum")
```

```
barplot(formula = ag$runs ~ ag$type, xlab = "Type", ylab = "Total Runs", col = "orange", axis.lty = 4, main = "Bar Plot by Aggregation (runs=166)")
```

Two new functions are introduced in this Bar Plot configuration. The *aggregate* function computes subsets of data. The *barplot* function provides the means by which to visualize the aggregated results.

Bar Plot by Aggregation (runs=166)

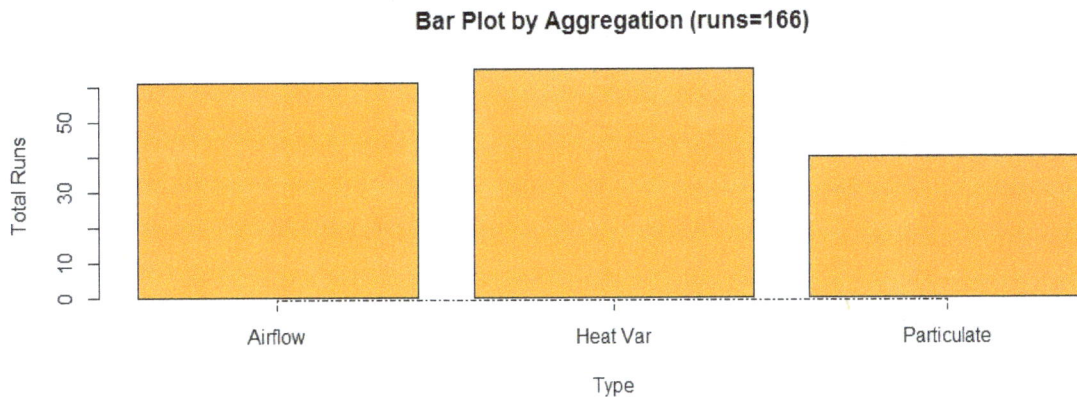

In this Bar Plot example, the runs are first subgrouped by type then added together. The results are then converted into a Bar Plot. Bar Plots using aggregation generally consist of two lines of code. The first code line creates an aggregated data frame in which a uniquely peculiar formula syntax is applied. The formula argument in the barplot function uses a similar code syntax. Critical arguments used in the *aggregate* function are described below:

- **formula** – Establishes the computational relationship between the quantitative vector and the categorical vector against which the data is modeled. The tilde (~) is used to denote the phrases, "is a function of", "is dependent on", or "is a response to." So, the aggregate function's formula argument can literally be interpreted as "run values are a function of the type", "run values are dependent on the type", or "run values are a response to the type." When using the formula argument to generate a Bar Plot, vectors containing quantitative values are referenced first, followed by a ~, then concluding with a reference to the qualitative vector. The parameter order is critical.
- **data** – The dataset used to generate aggregated output.
- **FUN** – The arithmetic function used to define the aggregation. Acceptable arguments include but are not limited to max, mean, median, min, prod, range, sd, sum, and var. The arithmetic function must be enclosed with quotation marks.

To get the total number of runs by measurement type, execute the ag object by typing "ag" in the Console window then press the Enter key:

```
ag
          type runs
1      Airflow   61
2     Heat Var   65
3  Particulate   40
```

The *barplot* function uses an aggregated dataframe to create the plot. Critical arguments used in the *barplot* function are described below:

- **formula** – This argument is similar in code syntax construction to the formula argument provided in the aggregate function. However, instead of referencing only the vector name, the syntax requires the dataset name to be included. The code syntax for this argument is *dataset_name$vector_name*.

- **xlab** – The label name of the plot's x axis.
- **ylab** – The label name of the plot's y axis.
- **col** – The color(s) used for the rectangular bars.
- **axis.lty** – This graphic parameter, referred to as the Axis Line Type, applies directly to the styling of the Bar plot's x axis. There are a total of 6 axis styles from which to choose. The argument is generally encoded using a numeric value between 1 and 6. The bar axis for this plot is suppressed by default requiring this parameter to be specifically called.
- **main** – Sets the Bar Plot's title. The total number of samples comprising the dataset being plotted should also be included in the title. This is encoded as "n=[total sample number]." For the purposes of simplicity and clarity, *n=* is replaced by *runs=*.

Finally, to retrieve the total number of runs captured in the aggregated Bar Plot, use the *sum* function against the runs vector in the ag dataset:

sum(ag$runs)

[1] 166

Box Plot

A Box Plot is a powerful visualization used in statistics. It is used to help understand the shape of a numeric distribution of data through the perspective of quartiles. Quartile characteristics of a distribution are captured including the minimum, maximum, IQR, and median range. Quartile data are presented through a colorized rectangle. Thus, the term "Box Plot." Outlier values, if they exist, are plotted as circles. The vertical dotted line is referred to as the "whisker." A Box Plot diagram describing each of these elements is illustrated below:

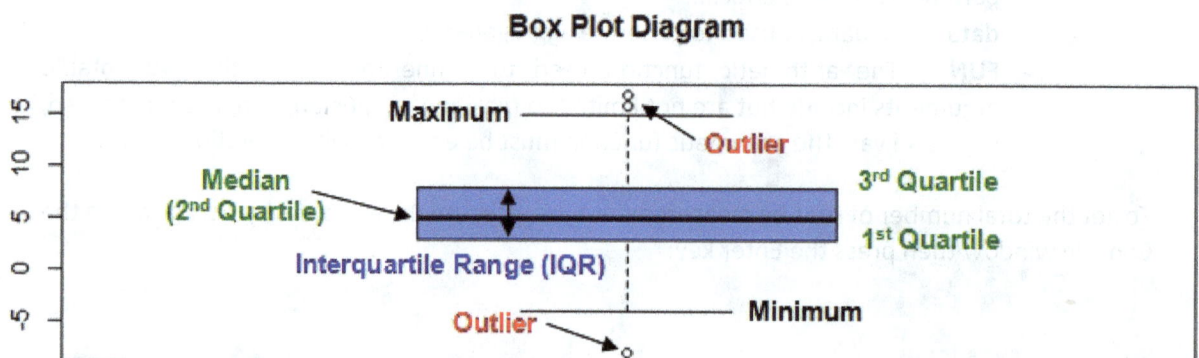

Box Plot Diagram

Like the Bar Plot, Box Plots can be created by *itemization* or by *aggregation*. Using the *wid* dataset, a Box Plot that captures the quartile distribution of all the average measurements by *itemization* is presented below. To generate this plot, a new function called *boxplot* is introduced.

boxplot(x = wid$measure, ylab = "Avg Measure", col = "mediumslateblue", main = "Wid Box Plot by Measurment (n=50)")

Wid Box Plot by Measurment (n=50)

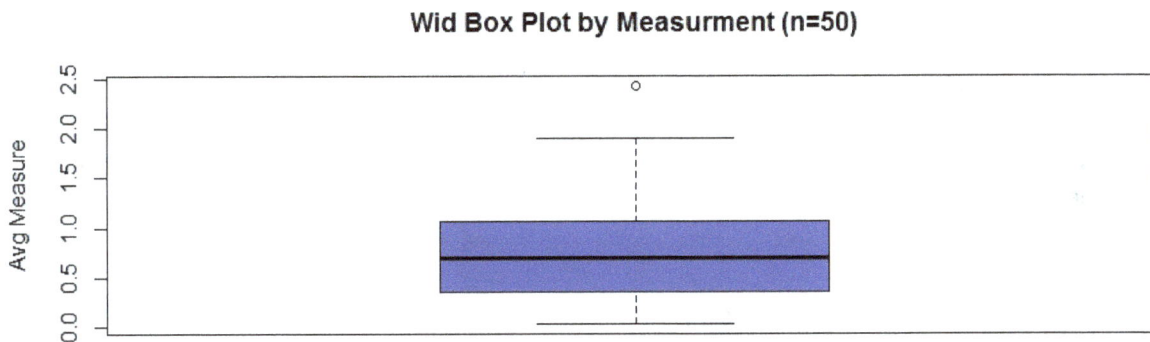

What this plot shows is that the interquartile range is located towards the lower end of the distribution. Another striking feature of the Box Plot is that an outlier exists above the upper cross-hatch. To retrieve the actual measurements representing the quartile characteristics of the Box Plot, an R function called *boxplot.stats* is introduced:

boxplot.stats(wid$measure)

$stats
[1] 0.03206127   0.35419883   0.69879916   1.06636311   1.90602869

$n
[1] 50

$conf
[1] 0.5396691   0.8579292

$out
[1] 2.428709

Output from the *boxplot.stats* function provides four critical pieces of information about the Box Plot called *stats*, *n*, *conf*, and *out*.

Box Plot Quartile Statistics

The *$stats* line item returns the principal quartile characteristics of the distribution. These characteristics are as follows:

- **Minimum Value** – The first value returns the minimum value of the distribution at 0.03206127. Values smaller than the minimum value are considered outliers.
- **First Quartile** – The second value returns the 25th percentile or the lower hinge at 0.35419883
- **Median** – The third value returns the median or the 50th percentile of the plot at 0.69879916
- **Third Quartile** – The fourth value returns the 75th percentile or the upper hinge at 1.06636311
- **Maximum Value** – The fifth value returns the maximum value of the distribution at 1.90602869. Values larger than the maximum value are considered outliers.

Total Records

The *$n* line item returns the total number of records, also referred to as observations, comprising the sample. Using the *wid* dataset as the sample, that number is 50. This number should be added to the plot's title to identify the sample size.

Confidence Interval

The *$conf* line item returns statistically significant quartile values adjusted for sample size. The values 0.5396691 and 0.8579292 represent plausible estimations of the first and third quartiles respectively, with a 95% confidence level. The purpose of this metric is to define a highly credible interquartile range within a distribution that truly exists. The real value of this metric emerges when small and large distributions of similarly defined data are compared.

Outliers

The *$out* line item returns outlier values identified within a distribution. Outliers lie beyond the statistical extremes of the distribution's minimum and maximum values. One outlier was identified within the *wid* dataset at 2.428709. Examining additional attributes of the records to which outlier values belong is sure to reveal additional and interesting information about the dataset.

Box Plot by Aggregation

While a Box Plot created by *itemization* provides information about the distribution of a numeric vector, it does not include distributions by category. If data is categorized in a dataset, then plotting by subgroup is plausible. When possible, plotting by both *itemization* and *aggregation* should be considered. Using the *wid* dataset, a Box Plot by *aggregation* is created by applying the following code:

```
boxplot(formula = wid$measure ~ wid$type, xlab = "Type", ylab = "Avg Measure", col = c("red", "blue", "yellow"), main = "Wid Box Plot by Type Aggregation Using Average Measure (n=50)", las = 1)
```

The output produces the following aggregated Box Plot:

Wid Box Plot by Type Aggregation Using Average Measure (n=50)

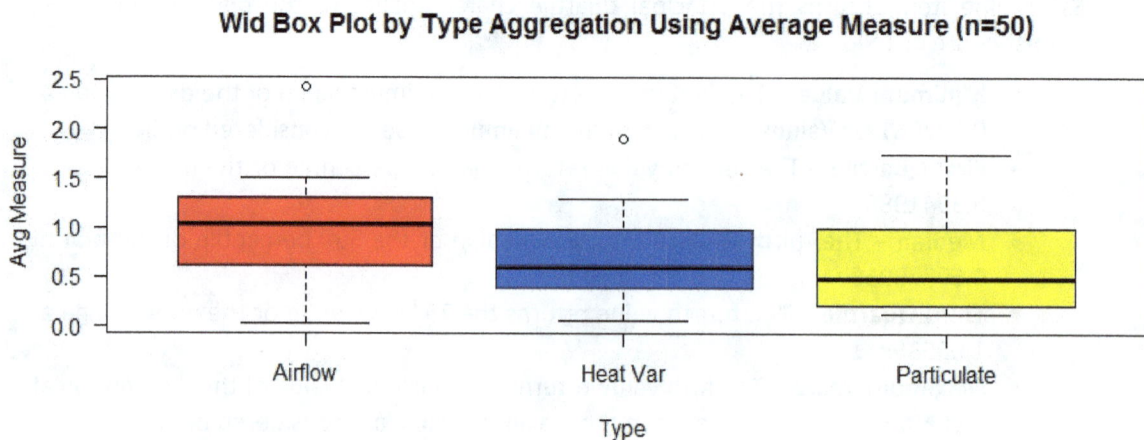

What should immediately be noticeable is the pattern with which the code in these exercises has been constructed. Arguments such as *formula*, *xlab*, *ylab*, *main*, and *col* are commonly represented in R plotting functions. With rare exception, the meaning of these arguments as they are applied within a plot function are the same. There is, however, an argument that has been added to the Box Plot aggregation code not used in previous plot functions. It is the *las* argument. This argument controls the direction of the tick mark labels on the x and y axes. For example, compare the y axis labels of both the itemized and aggregated Box Plots. In the *Box Plot by Measurement*, the y axis labels run vertically. In the *Box Plot by Type Aggregation*, the y axis labels are plotted laterally. Acceptable *las* argument values range from 1 to 3.

So, what kind of information does an aggregated Box Plot convey? To get the actual values represented by this Box Plot visualization, the *boxplot* function needs to be reformulated. The *boxplot* function can be used to both visualize and extract critical data defining its distribution. It is a clever convenience provided in R. To capture actual data values from the Box Plot, apply the following code:

```
boxag = boxplot(formula = wid$measure ~ wid$type, plot = FALSE)
```

This code captures, converts, then stores the Box Plot data into a variable called boxag. An argument called *plot* is set to FALSE to prohibit a Box Plot from being generated. To view the results, type "boxag" in the Console window then press the Enter key:

```
boxag
$stats
          [,1]        [,2]        [,3]
[1,] 0.03206127  0.05916393  0.1264153
[2,] 0.60888995  0.37896817  0.2163394
[3,] 1.03438688  0.60035202  0.4845674
[4,] 1.30485091  0.97386432  0.9891569
[5,] 1.50565542  1.28610701  1.7412872

$n
[1] 16   22   12

$conf
         [,1]        [,2]        [,3]
[1,] 0.7594823   0.399957    0.1320801
[2,] 1.3092915   0.800747    0.8370546

$out
[1] 2.428709   1.906029

$group
[1] 1 2

$names
[1] "Airflow"   "Heat Var"   "Particulate"
```

Like the output provided by the *boxplot.stats* function, the quartile results from the *boxplot* function include box plot quartile statistics, number of records, confidence intervals, and outliers for each category under analysis. However, two data dimensions called *$group* and *$names* are added to a Box Plot aggregation analysis.

Outlier Groups

The *$group* line item returns the categories, by number, to which any outliers belong. For example, in the *wid* dataset, there are three widget-type categories under analysis: Airflow (1), Heat Var (2), and Particulate (3). The results indicate that outliers have been identified in groups 1 and 2.

Names

The $names line item returns a listing of the categories defined by the second parameter of the *boxplot* function's formula argument. Originating from the *wid* dataset's type vector, the categories are Airflow, Heat Var, and Particulate.

Fan Plot

The Fan Plot possesses visual characteristics similar to a Pie Chart. However, the Fan Plot is both more informative and compelling in its visualization. When a Pie Chart is being considered as a graphic visualization, a Fan Plot, instead[17] should be used. A Fan Plot is a plot in which numeric values, proportioned either by percentage or frequency, represent the arcs of overlapping categories. In a fan plot, the slices are arranged from the smallest to the largest slice in a way where each slice overlaps. The radii are constructed in a way that shows each slice. This visual arrangement creates a "fan" look. Like a Pie Chart, the number of categories used to create a Fan Plot are limited.[18] To create a Fan Plot, an R package called *plotrix* must first be installed. Once installed, the package is then loaded for use in the session. The package only needs to be installed once.

install.packages("plotrix")

library(plotrix)

The function used to create a Fan Plot is called *fan.plot*. There are several arguments comprising this function. However, the arguments minimally required to produce the Fan Plot are as follows:

- A numeric vector called *x*
- Plot labels called *labels*
- A plot title referred to as *main*

---

[17]This comment applies specifically to Pie Charts being considered as graphical visualizations in R. It would not apply, for example, to Pie Charts used in interactive reporting where business intelligence dashboards are being used.

[18]Modeling the data visually will determine whether the number of categories being considered is appropriate for a Fan Plot.

Colors can be customized by using the function's *col* argument. However, the colors provided by default should be sufficient. In this exercise, a Fan Plot will be created by using data provided from the *Box Plot by Type Aggregation* example.

plot.new()

dev.new()

labs = paste(boxag$names, boxag$n)

fan.plot(x = boxag$n, labels = labs, main = "Widget Performance Type Fan Plot (n=50)")

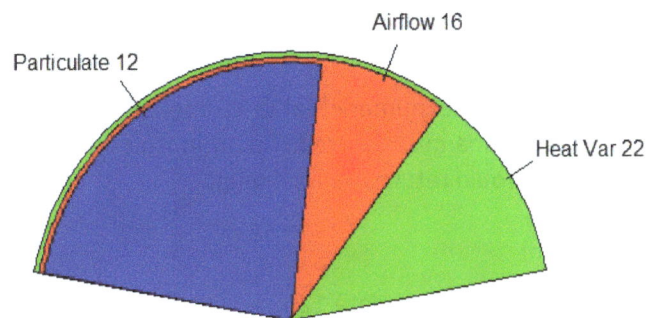

Code Explanation

The *plot.new* and *dev.new* code lines must be added to this solution or else the plot will be visually truncated when generated in RStudio's classic plot window. When executed, these code functions symbiotically create a new plot window within which a new plot device is used to correctly draw the Fan Plot image. The example below illustrates how this new plot window device is displayed.

New Plot Device Window Used to Create a Fan Plot

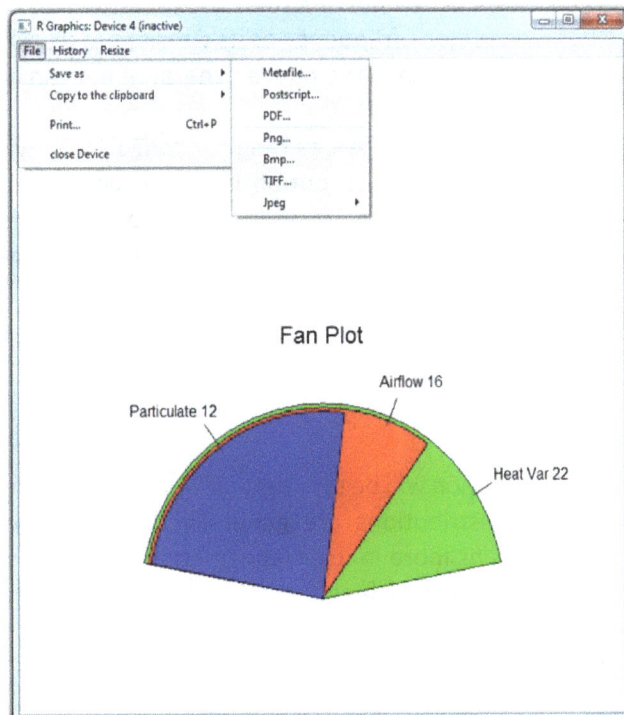

As the image shows, the Fan Plot can be saved to a variety of file formats. The next line of code creates both the labels and values used in the plot. To generate these labels, a new function is introduced called *paste*. It is particularly suitable for plot labeling and customization. This function provides the means by which data strings can easily be conjoined. The technical term used for this action is called "string concatenation." Using the *boxag* variable from the previous exercise, Fan Plot labels can be efficiently constructed.

The *paste* function applies two arguments to define the plot's label format. These arguments are described as follows:

- **First Argument:** Creating plot labels linearly from left to right, label names are first extracted from the *names* vector contained in the boxag variable.
- **Second Argument:** Numeric values are extracted from the *n* vector of the boxag variable

A space between the label and the numeric value is implied and processed accordingly. In addition, there are no argument names required. Taken in isolation, the output from the code that generates the plot labels would return the following:

paste(boxag$names, boxag$n)

[1] "Airflow 16"   "Heat Var 22"   "Particulate 12"

Putting it all together, the final line of code generates the Fan Plot.

Histogram

The Histogram plot is one of the most common graphical visualizations used in statistics. Similar to a Bar Plot, a Histogram displays the distribution of a numeric vector by dividing the range of its values into a specified number of bins. The bins are then distributed across the plot's x-axis. Values representing either the numeric vector's frequency or its density are displayed on the y-axis. The number of bins is generally optimized when the plot is generated but can also be customized.

In this exercise, a numeric vector of 1,000 values ranging from 1 to 50 will be used to create a symmetrical distribution. A symmetrical distribution is one in which both the mean and the median measurements are nearly equivalent. The following code creates a sample distribution of numeric values that are symmetrical in form. For the purpose of demonstration, the *set.seed* function is included to provide distributive reproducibility.

set.seed(seed = 21092, kind = "Mersenne-Twister")

sym_shape = sample(x = 1:50, size = 1000, replace = TRUE)

How do you know that this distribution will be symmetrical? Plotting a Histogram will clearly show the distributive arrangement. Distributions are generally defined in one of three ways. A distribution of numeric values leans more in its arrangement towards being either symmetrical, skewed right, or skewed left. By applying a function in R called *hist* to the sym_shape variable, its distribution can be visually observed through a Histogram.

hist(x = sym_shape, main = "Histogram with Symmetrical Distribution", col = "yellow", xlab = "Value", las = 1)

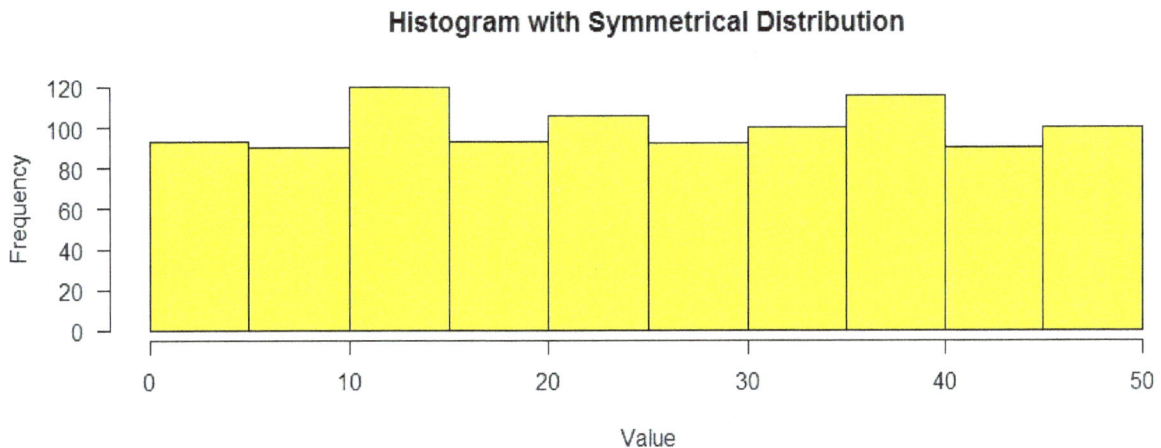

Histogram with Symmetrical Distribution

Consistent with its distribution, the Histogram shows the bins to be mostly symmetrical in height. To better understand the metadata associated with this plot, the code must be reformulated and stored in a variable.

hgmeta = hist(x = sym_shape, plot = FALSE)

The *hgmeta* variable, interpreted as "a histogram of metadata," provides detailed information about the distributive characteristics of the sym_shape numeric vector. To view the results, type "hgmeta" in the Console window then press the Enter key. The results are provided as follows:

hgmeta

$breaks
[1] 0 5 10 15 20 25 30 35 40 45 50

$counts
[1] 93 90 120 93 106 92 100 116 90 100

$density
[1] 0.0186 0.0180 0.0240 0.0186 0.0212 0.0184 0.0200 0.0232 0.0180 0.0200

$mids
[1] 2.5 7.5 12.5 17.5 22.5 27.5 32.5 37.5 42.5 47.5

$xname
[1] "sym_shape"

$equidist
[1] TRUE

attr(,"class")
[1] "histogram"

Interpreting Histogram Metadata

Each of the six line items captured in the hgmeta data object describe a different distributive characteristic about the sym_shape vector. A brief description of each line item is provided below:

- **$breaks:** Defines the numeric boundaries separating the bins. For example, the Histogram for the sym_shape vector shows 11 breaks across 10 bins.
- **$counts:** Presents the total counts of the numeric values contained within each bin. For example, in the first bin, there are a total of 93 values captured. In the third bin, there are 120. The cumulative total is 1,000, the total number of values comprising the sym_shape vector.
- **$density:** These values are referred to as the Histogram's *frequency density*. Frequency density represents a computation of the area defined by each bin. For example, the frequency density for both the first and fourth bins is .0186. This value is calculated by dividing the frequency by the group width (.93 ÷ 50).
- **$mids:** These values represent bin-width midpoints. For example, the second bin contains values from 5 to 10. The bin's midpoint would therefore be 7.5.
- **$xname:** Represents the name of the numeric variable being examined. In this example, the variable name is sym_shape.
- **$equidist:** Indicates whether the distances between the bin breaks are the same. The result is a binary indicator returning either a True or False value.

The *attr(,"class")* property of the hgmeta data object is not a retrievable metadata line item per se. It provides supporting information regarding object encoding. This line item indicates that the hgmeta data object is encoded as class histogram. To verify this information, use the class function.

```
class(hgmeta)
```

```
[1] "histogram"
```

Skewed Distributions

To show the contrast of various distribution shapes represented in a Histogram plot, the next two plots will illustrate skewed distributions. A skewed distribution is one in which the mean is either greater or less than the median value of the distributive range. More specifically, if the mean is greater than the median, the distribution is said to be "skewed right." If the mean is less then the median, the distribution is "skewed left."

Similar to the previous exercise, the code below creates two numeric vectors each of which contain 1,000 values ranging from 1 to 50. However, these vectors show different distributions of data skewed both left and right. The results are stored in variables called skw_left and skw_right respectively.

```
set.seed(seed = 711, kind = "Mersenne-Twister")

par(mfrow = c(1, 2))

skw_left = sample(x = 1:50, size = 1000, replace = TRUE, prob = 1:50)

skw_right = sample(x = 1:50, size = 1000, replace = TRUE, prob = 50:1)
```

```
hist(x = skw_left, main = "Histogram with Skewed-Left Distribution", col = "aquamarine4", xlab =
"Value", las = 1)

hist(x = skw_right, main = "Histogram with Skewed-Right Distribution", col = "cornsilk3", xlab =
"Value", las = 1)
```

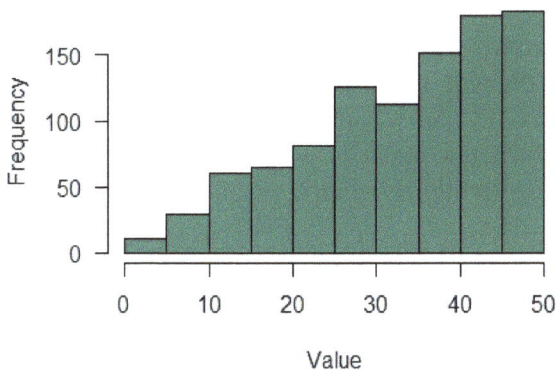

Histogram with Skewed-Left Distribution

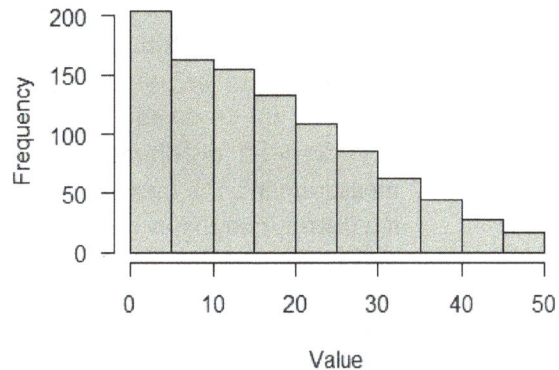

Histogram with Skewed-Right Distribution

These six lines of code accomplish several things. A brief explanation of each code line is provided below:

- **1st Code Line:** Used to provide distributive reproducibility. Ensures that both the data and results shown in the exercise will match those of an independent effort when correctly followed.
- **2nd Code Line:** Introduces a new function called *par*. This function is used to capture more than one plot on a single plot surface in RStudio. The argument mfrow defines the structural layout for multiple plots. The argument is short for *multiple figures use row-wise*. The configuration "c(1, 2)" means to plot all forthcoming plots in a one-row-two-column layout. Conversely, a configuration of "c(2, 1)" would be a two-row-one-column layout.
- **3rd Code Line:** This code generates a distribution of values that are skewed left.
- **4th Code Line**: This code generates a distribution of values that are skewed right.
- **5th Code Line:** Working in conjunction with the second code line, generates the first part of the plot as a skewed-left distributive Histogram.
- **6th Code Line:** Working in conjunction with the second code line, generates the second part of the plot as a skewed-right distributive Histogram.

Notice that the *prob* argument in the third and fourth code lines control for the probabilities of each value used within the distribution. This argument shapes the distribution types provided in this exercise. In the symmetrical distribution, the *prob* argument is not used.

Finally, to verify the statistical relationship between the shape of a distribution and its measure of centrality, both the mean and median values of the distribution are computed. The mean and median values for each distribution including *sym_shape*, *skw_left*, and *skw_right* are provided below:

mean(sym_shape)	mean(skw_left)	mean(skw_right)
[1] 25.619	[1] 33.643	[1] 16.927
median(sym_shape)	median(skw_left)	median(skw_right)
[1] 25	[1] 36	[1] 15

Based on these computations and the Histograms used to plot the data distributions, the following conclusions can be drawn:

- A symmetrical distribution is one in which both the mean and the median measurements are nearly equivalent. The *sym_shape* distribution measures of centrality are nearly equivalent, making it a symmetrical distribution.
- A skewed-left distribution is one in which the mean is less than the median measurement. The mean of the *skw_left* distribution is less than its median, making it a distribution that is skewed left.
- A skewed-right distribution is one in which the mean is greater than the median measurement. The mean of the *skw_right* distribution is greater than its median, making it a distribution that is skewed right.

Without the use of statistical computations that support visualizations like Histograms and Box Plots, it would be very difficult to successfully evaluate the distributive properties of numeric data.

Line Plot

In the graphical visualization of data, a Line Plot is one of the most popular and easiest plot types to understand. A Line Plot displays information as a series of points called "markers" through which a line is connected. In addition, a Line Plot can show one or more linear dimensions of data. Each linear dimension is separately defined and is represented from left to right across the plot region. The most effective and visually appealing way to generate a Line Plot in R is to use the R base *plot* function. This function offers a unique balance between visual aesthetics and code economy. In fact, the *plot* function is an intelligent graphical function used to connect data with an assortment of visualizations. The plotting flexibility of this function is truly remarkable. Its plotting capabilities include Scatter Plots, Bar Plots, Linear Regression Models, Line Plots, Box Plots, Mosaic Plots, and Dendrograms. The *plot* function is characterized as "intelligent" due to its incisive code-parsing capability. When applying this function, code syntax structure and data type encodings determine the plot type returned.

The key to developing advanced plots in R is that each plot part, typically represented by a successive line of code, serves as a separate plot "layer." Each layer is then superimposed on the plot palette. The strategic accretion of these layers is then used to generate a persuasively compelling plot.

In this exercise, two Line Plot types, consisting of multiple layers will be constructed. One Line Plot is unidimensional in form, consisting of a single linear dimension. The other Line Plot is multi-dimensional in its construction. The objective of this exercise is to show how to strategically build and use plot parts to create a commanding Line Plot.

Simulating A Dataset

Consistent with previous exercises outlined in this chapter, the first step in creating a plot in R is to establish a dataset. In this example, a simulated dataset will be used that captures the academic performance scores of a student recorded over a three year school period. A total of 15 test scores were recorded for each school year from the ninth through the eleventh grades. The simulated dataset, called *test_perf*, is provided below:

```
set.seed(seed = 95, kind = "Mersenne-Twister")

Grade9 = sample(x = 70:80, size = 15, replace = TRUE)

Grade10 = sample(x = 80:100, size = 15, replace = TRUE)

Grade11 = sample(x = 70:100, size = 15, replace = TRUE)

test_perf = data.frame(Grade9, Grade10, Grade11)
```

To review the first six records of the *test_perf* dataset, use the head function:

```
head(test_perf)
```

```
  Grade9 Grade10 Grade11
1     76      85     100
2     80      97     100
3     77      98      88
4     78      95      83
5     77      99     100
6     72     100      98
```

Unidimensional Line Plot

Once a dataset has been established, the next step is to select a plot type. In this example, a unidimensional Line Plot will be used to capture the student's eleventh grade school year performance scores. After the data and the plot type have both been determined, plot building can begin.

The first code line establishes the Line Plot's structural baseline. This baseline can be referred to as *layer 1* of the plot. *Layer 1* includes plot characteristics such as the interval against which the plot is being measured, actual performance results, x and y label names, line type and color, and the plot's title.

```
plot(x = 1:15, y = test_perf$Grade11, xlab = "Test Interval", ylab = "Performance Score", main = "Student's 11th Grade Academic Performance Line Plot (n=15)", las = 1, col = "blue", type = "l", lwd = 2)
```

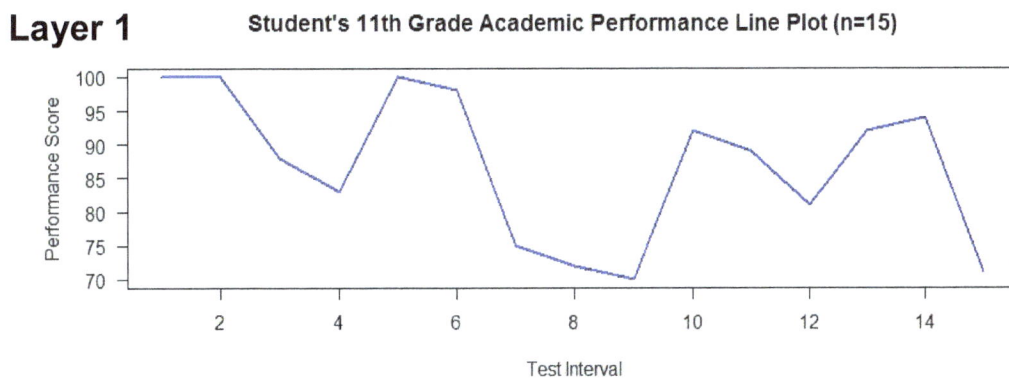

Layer 1 — Student's 11th Grade Academic Performance Line Plot (n=15)

In using the *plot* function to create a Line Plot, two new arguments are introduced. Located towards the end of the function, these arguments are *type* and *lwd*. The *type* argument determines how the data dimension being plotted is manifested in the plot. The most common argument values are "l" for line, "p" for points, "b" for both points and lines, or "s" for a stair-step style Line Plot. The stair-step style Line Plot uses right angles exclusively to create the plot. The *lwd* argument controls the plot's line width. Reasonable values for this argument range from 1 to 10. The larger the value, the thicker the line.

In theory, *Layer 1* provides enough visual information to capture the essence of a Line Plot but is incomplete. To develop a compelling Line Plot requires additional layers of visual information. *Layer 2* adds colorized points called markers. Markers are independent of line color. The markers are solid circles colorized in red. In this example, 15 markers represent the actual values extracted from the Grade11 vector of the test_perf dataset. As a newly introduced function, the *points* function defines *Layer 2* of the Line Plot.

points(x = test_perf$Grade11, pch = 16, col = "red", cex = 1.2)

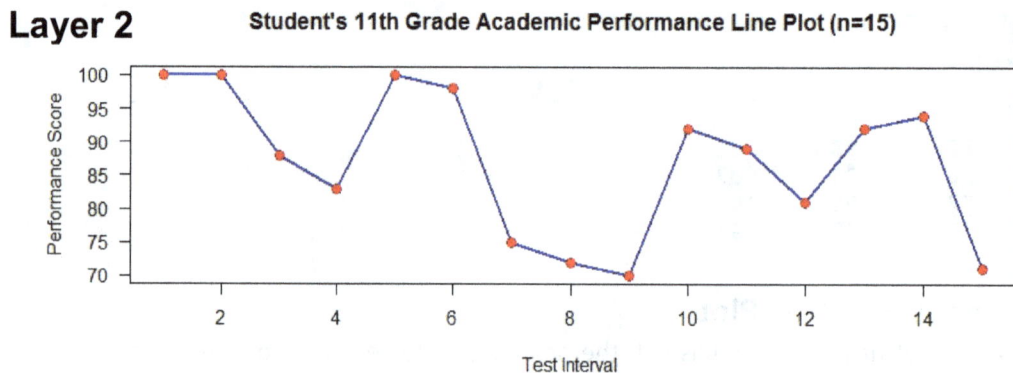

The primary arguments used in the *points* function are as follows:

- **x:** This argument references the numeric vector used to define the plot's data points, or markers.
- **pch:** This argument is an abbreviated term for "plot character." It uses a numeric value to define the marker type.
- **col:** This argument defines the marker color.
- **cex:** Controls the marker size in values of one-tenth increments. For example, 1.3, 1.4, 1.5, etc.

To expand on the various options available as argument values used in plot development, a general reference listing of markers, symbols, colors, and lines is available in R. This reference listing is accessible as an independent plot by executing the *PlotPar* function found in the *DescToolsAddIns* package. To access this reference listing, an R package called *DescToolsAddIns* must first be installed, then loaded.

```
install.packages("DescToolsAddIns")

library(DescToolsAddIns)

PlotPar()
```

Argument Options Available in R Plot Development

The third layer adds precision to the Line Plot. Applying a new function called *text*, marker labels are added with one line of code.

```
text(x = test_perf$Grade11, labels = test_perf$Grade11, pos = 2, cex = .8, col = "red")
```

The primary arguments used in the *text* function are as follows:

- **x:** This argument references the numeric vector used to define the plot's data points, or markers.
- **labels:** This argument defines the marker's label names.
- **pos:** This argument controls the placement of the text labels relative to the markers. Acceptable values range from 1 to 4.
    - 1 = Label below the marker
    - 2 = Label to the left of the marker
    - 3 = Label above the marker
    - 4 = Label to the right of the marker

- **cex:** Controls the size of the text label in values of one-tenth increments. For example, .6, .7, .8, etc.
- **col:** This argument defines the text label color.

Layer 3

Student's 11th Grade Academic Performance Line Plot (n=15)

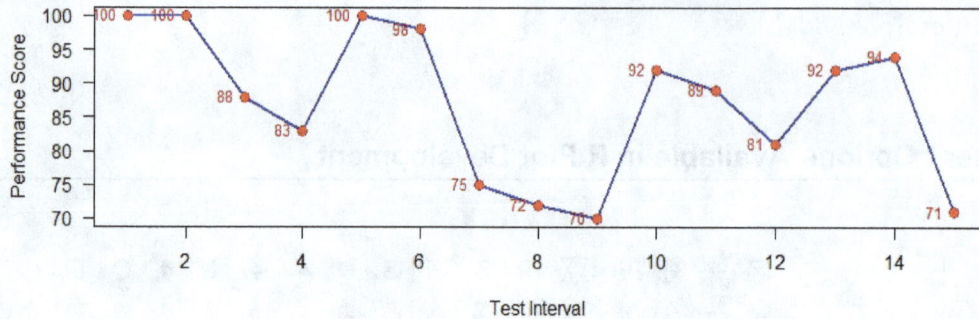

The fourth and final layer of the unidimensional Line Plot introduces a new function that adds a lightly colored plot grid. A plot grid adds visual coordination to the plot. It blends the x and y axis coordinates together in a way that improves plot acuity. Coordinate grid color, type, and width can be controlled by the *col*, *lty*, and *lwd* arguments respectively.

grid()

Layer 4

Student's 11th Grade Academic Performance Line Plot (n=15)

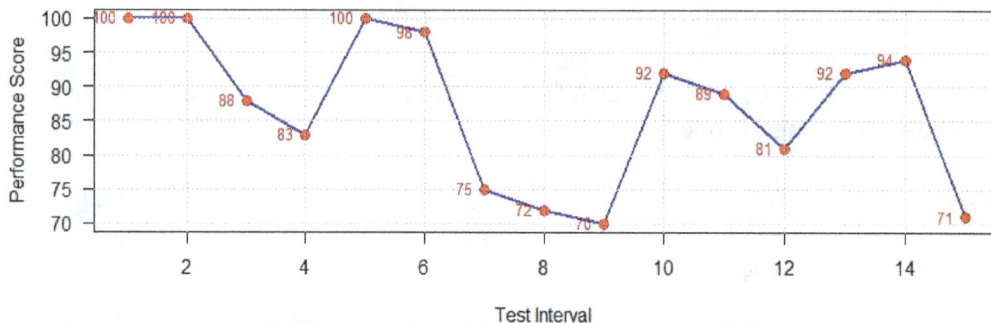

When developing multi-layer plots, any mistake made in the construction of a layer will require the entire plot to be re-built. It is always a good practice to type each layer of code in the Script Editor window as the plot is being developed. If an error is made in the code or the plot results don't meet visual expectations, the code in the Script Editor window can be easily modified. Once modified, the code from the Script Editor can then be copied and pasted for re-execution in the Console window. Alternatively, pressing the ↑ Arrow key on the keyboard in the Console window will scroll through the history of code previously executed one line at a time. Accessing code in this manner prevents having to re-type commands.

Multi-Dimensional Line Plot

For numeric data dimensions that share a common measure, a multi-dimensional Line Plot offers a suitable plot solution. This plot type displays multiple linear dimensions of data on a single plot palette. Care should be taken to limit the number of linear dimensions used in a multi-dimensional Line Plot. Plotting excessive linear dimensions reduces the plot's visual acuity and will impair its effectiveness. Conversely, generating plots with limited or incomplete plot parts can be confusing to understand, and in some cases, misleading.

In this exercise, a multi-dimensional Line Plot will be generated. Using the *test_perf* dataset from the previous exercise, this plot will show the dimensional comparisons in a student's academic performance from the ninth through the eleventh grades. Layer plotting will be used.

The first step is to assess the range of each numeric vector to be plotted. The objective is to identify the vector with the widest range. This vector, by way of its linear dimension, will be plotted ***first***. It is important to plot the linear dimension with the widest range first so that other plot dimensions are not subsequently truncated. Wider ranged dimensions will be truncated if smaller ranged dimensions precede them during the plotting process. The multi-dimensional Line Plot below provides an example showing the effects of truncation. The range of the green line is being constrained by the smaller ranged blue line which was plotted first. Due to its wider range, the green line shows breaks in its linear dimension, impeding the ability to be completely displayed.

Truncated Multi-Dimensional Line Plot

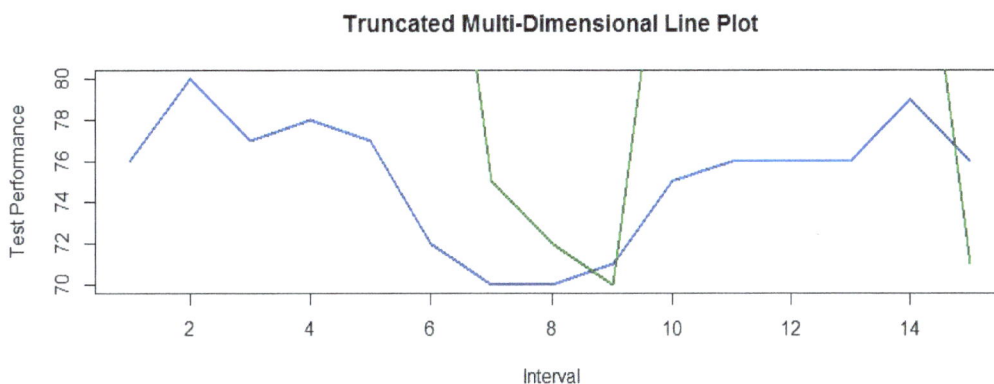

To determine which dimension should be plotted first, calculate the range of each numeric vector that will be used as a line segment in the plot. This can be achieved by using the *range* function:

range(test_perf$Grade9)
[1] 70 80

range(test_perf$Grade10)
[1] 81 100

range(test_perf$Grade11)
[1] 70 100

Based on the results, the Grade11 vector has the widest range with a minimum value of 70 and a maximum value of 100. So, this dimension will be plotted first. The order of the remaining dimensions is inconsequential, unless an order of some kind can be organically captured. For example, in this exercise, the Grade11 linear dimension will be plotted first. To preserve sequential order, the Grade10 and Grade9 vectors will follow. However, if Grade10 had been the vector with the widest range, plotting by sequential order would not be possible. Capturing order is important in a multi-dimensional Line Plot due to its impact on the plot's legend. A multi-dimensional Line Plot with ordered dimensions is much easier to follow and understand.

Layer 1 of the multi-dimensional Line Plot generates the first linear dimension. This line segment is colorized purple.

```
plot(x = 1:15, y = test_perf$Grade11, xlab = "Time Interval", ylab = "Test Performance", main = "Student Performance Test Results Line Plot (n=15)", las = 1, col = "blueviolet", type = "l", lwd = 2)
```

Layer 1

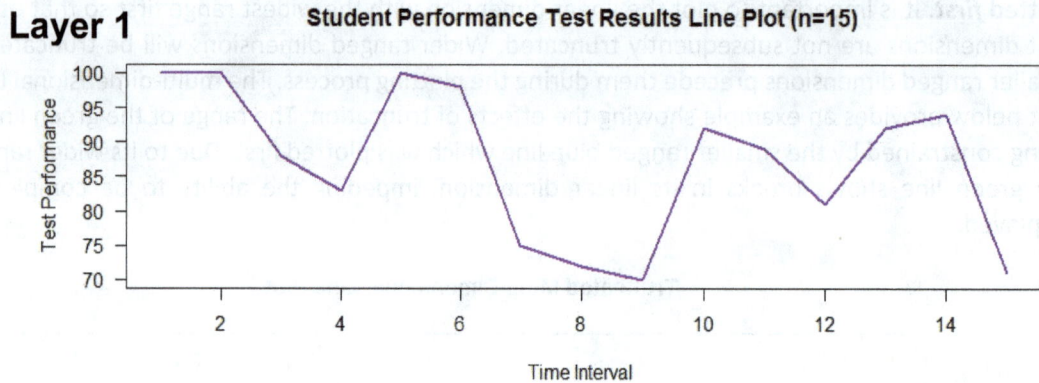

Layer 2 adds colorized markers. Due to multi-dimensional plotting, marker sizes for each layer should be a little larger than those provided in a unidimensional Line Plot.

```
points(x = test_perf$Grade11, pch = 16, col = "blueviolet", cex = 1.2)
```

Layer 2

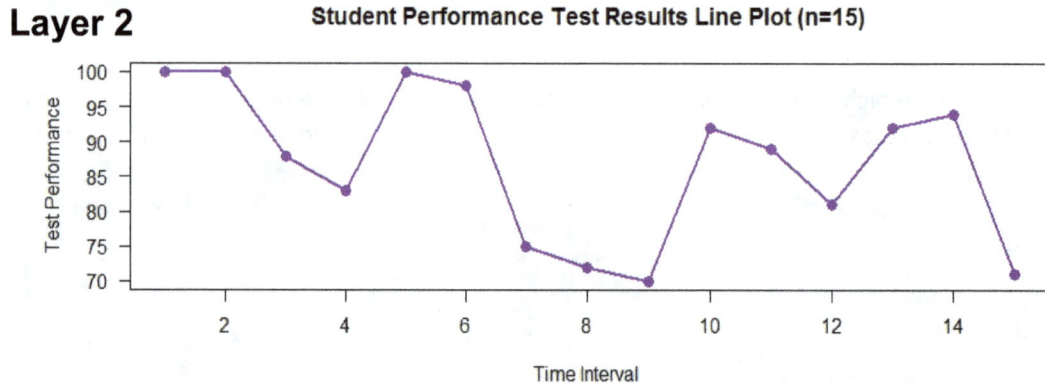

Layer 3 adds a new linear dimension to the plot. This line segment will represent the Grade10 vector. In this layer, a new function is introduced called *lines*. It is designed to be used with multi-dimensional Line Plots. Its arguments resemble those used in previous plot functions.

This layer comprises two lines of code. The first code line adds a linear dimension that is orange in color. The second code line adds the colorized markers:

```
lines(test_perf$Grade10, col = "orange", type = "l", lwd = 2)
```

```
points(test_perf$Grade10, pch = 16, col = "orange", cex = 1.2)
```

Layer 3

Student Performance Test Results Line Plot (n=15)

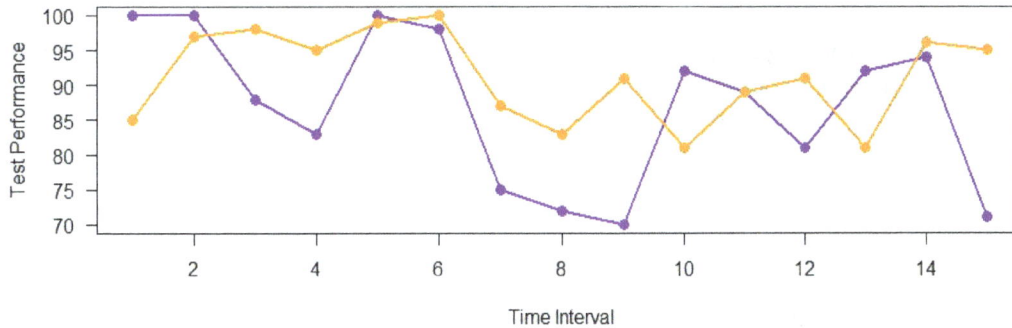

The final linear dimension of the plot adds the Grade9 vector. Colorized in green, code provided in *Layer 4* replicates *Layer 3* code in its syntax.

```
lines(test_perf$Grade9, col = "green4", type = "l", lwd = 2)

points(test_perf$Grade9, pch = 16, col = "green4", cex = 1.2)
```

Layer 4

Student Performance Test Results Line Plot (n=15)

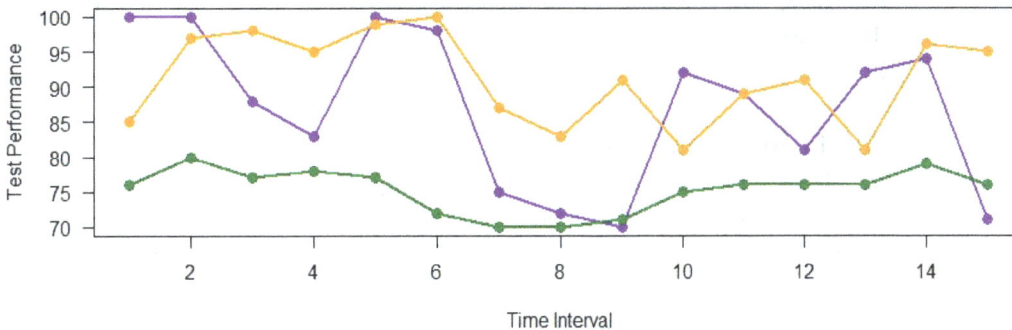

Multi-dimensional Line Plots are incomplete without a legend. The legend is a critical part of this plot type. It distinguishes linear dimensions by name. Without a legend supporting the plot, there would be no way to connect a linear dimension to its corresponding vector. *Layer 5* adds the final plot part to the Line Plot. By adding this plot part, this layer introduces a new function called *legend*.

```
legend(x = "topright", legend = rev(names(test_perf)),  title = "PERF", bg = "gray88", fill = c("blueviolet",
"orange", "green4"), cex = .6)
```

Layer 5

Student Performance Test Results Line Plot (n=15)

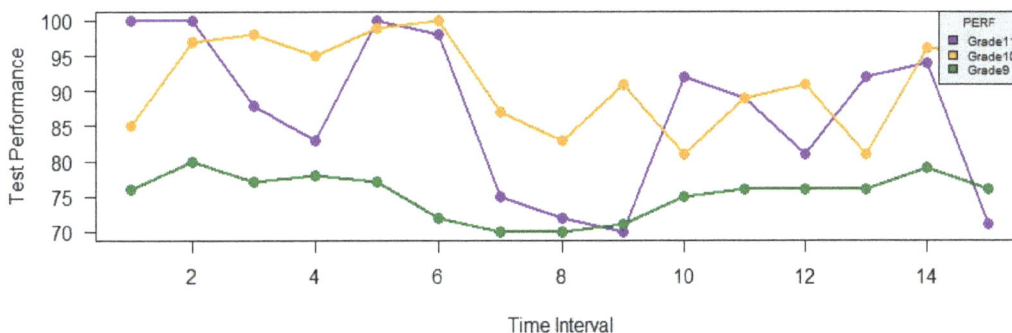

There are several arguments supporting the *legend* function. However, the arguments minimally required to add a legend as a plot part are as follows:

- **x:** This argument controls the placement of a legend on the plot. There are nine locations available on a plot palette where a legend can be positioned. The legend object should be positioned in a way that does not obstruct the ability to view the plot's line segments. In alphabetical order, these placement options are as follows:

  - bottom
  - bottomleft
  - bottomright
  - center
  - left
  - right
  - top
  - topleft
  - topright

- **legend:** Defines the names of the linear dimensions used in the legend. A new function is introduced in conjunction with the value of this argument called *rev*. This function reverses the order of vector elements from left-to-right to right-to-left. In this case, the *rev* function is used to realign the sequential order of the linear dimensions from their original position within the *test_perf* dataset. This ensures that both the line segment colors and vector names are correctly connected in the plot.

- **title:** Establishes the legend's title.

- **bg:** This argument determines the background color of the legend box. The legend box should be a color different from the plot's background color. A light gray color is recommended. This color allows the legend to be both recognizable and readable on the plot palette.

- **fill:** Defines the legend's box colors. These colors should match the line segment colors used in the plot.

- **cex:** This argument defines the size of the legend box proportional to its title, box colors, and dimension names. Values should be set in one-tenth increments. For example, .5, .6, .7, etc.

Notes

A total of seven lines of code comprising plot parts organized across five layers were used to create a multi-dimensional Line Plot. As this exercise demonstrates, there are many plot options available in R. Experiment with the possibilities. Using these exercises as a guideline, create a process model that best suits your plotting objectives.

Mosaic Plot

Of all the plots profiled, the Mosaic Plot is perhaps one of the most obscure. It is a plot with which most people are generally unfamiliar. One potential reason for the plot's lack of familiarity is that only qualitative data are used to generate the plot. Conversely, most plots used to analyze today's data apply at least one quantitative variable. However, Mosaic plots process data exclusively by groups. Another way to describe the Mosaic plot is that it is a cross-group analysis defined by a graphical visualization. A Mosaic Plot can provide remarkable insights into categorical data. In this exercise, a Mosaic plot will be used to visually capture a categorical analysis of qualitative data.

The Setup

A fictional banking institution called ACME Bank recently conducted a meeting regarding the state of its customer base. Upon conclusion of the meeting, a primary question emerged. Given the bank's active position in wanting to serve a customer base that exemplifies a high degree of credit worthiness, what measures can be taken to achieve this objective? A supplemental question, more specific to the objective was posed. How many customer accounts would need to be removed from the system to meet this objective? Many cascading questions naturally emerge from broad inquiries of this type. Before these questions can be answered, an initial assessment of the state of ACME Bank's customer base must first be conducted. A baseline that captures a reasonable set of statistically significant characteristics about the bank's customer base can be used to answer these, and other supplemental questions. For this assessment, a Mosaic Plot will be used to establish the current state of ACME Bank's customer base.

Simulating A Dataset

In this exercise, a Mosaic Plot will be used to statistically analyze the relationship between the credit scores of ACME Bank's customer base and the length of time that customer accounts have been open and active. A random sample of 2,500 customer credit scores and active account records will be used as the dataset. A simulated dataset, called *bnk*, is provided below:

```
set.seed(seed = 741776, kind = "Mersenne-Twister")

score = as.factor(sample(x = c("500", "501-800", "800"), size = 2500, replace = TRUE, prob = c(.65, .3, .05)))

actage = as.factor(sample(x = c("01-03", "04-12", "13-36", "37-84", "85+"), size = 2500, replace = TRUE, prob = c(.1, .15, .25, .3, .2)))

bnk = data.frame(score, actage)
```

To review the first six records of the *bnk* dataset, use the head function:

```
head(bnk)
    score actage
1 501-800  04-12
2     500  37-84
3     500    85+
4     500  04-12
5     500  37-84
6     500  13-36
```

A brief description of the vectors comprising the *bnk* dataset is as follows:

- **score** – A category of class factor into which a customer's credit score falls. There are three categories comprising this vector. Credit score categories include 500 or less (500), scores between 501 – 800 (501-800), and credit scores higher than 800 (800).

- **actage** – This category of class factor captures the number of months (age) that a customer account has been open or active. There are five categories defining this vector to include the following:

  - 01-03: Up to 1 month through 3 months
  - 04-12: From 4 to 12 months
  - 13-36: From 13 to 36 months
  - 37-84: From 37 to 84 months
  - 85+: 85 months or longer

Creating & Evaluating A Mosaic Plot

Before a Mosaic plot can be generated, a data object called a *two-way table* must be created. The *two-way table* computes a combination of cross-referenced subgroup counts from both the *score* and *actage* vectors. To capture these subgroup counts, a newly introduced function called *table* will be used. The output will be saved to a data object called tbl. The tbl object will then be used as the first argument in the *plot* function. The order of the categorical variables used in the *table* function is critical as it directly impacts the structure and display of the Mosaic Plot.

tbl = table(x = bnk$actage, y = bnk$score)

plot(x = tbl, xlab = "Act Open in Mths", ylab = "Credit Score", col = c("orange", "brown", "yellow"), main = "Account Age to Credit Score Assessment (n=2500)", cex = 1)

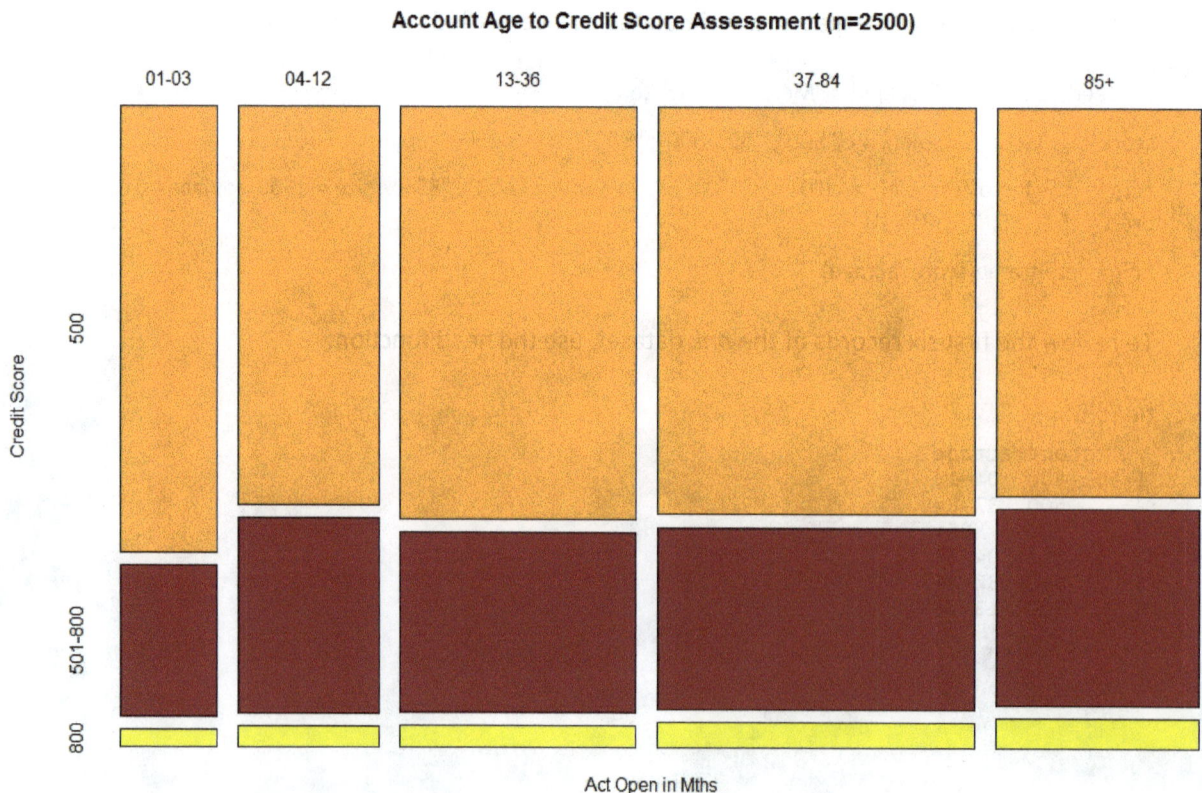

Account Age to Credit Score Assessment (n=2500)

The Mosaic plot provides insightful information. It shows the proportional relationships between the *score* and *actage* vectors. Understanding the structure of proportionality in the Mosaic plot is key to its interpretation. Simply put, the colors represent a percentage of credit scores by category. A credit score of 500 or less is orange, 501-800 is brown, and credit scores above 800 are colorized yellow. The width of each account age category is proportional in size to the total number of scores defined by its corresponding credit score category. In this plot, there are a total of five account aging categories on the x axis and three credit score categories on the y axis. Combining these proportions in a visual fashion presents a new perspective of ACME Bank's customer credit scores and active accounts.

It is important to conceptually understand the visual characteristics of a Mosaic Plot. However, it is equally important to know how to extract the plot's subgroup characteristics with specificity. What information specifically underlies the plot? The answer to this question is found in the tbl object. However, before viewing the tbl object's data, it must first be slightly modified. Two tbl object modifications are required to optimize the data used in this analysis. The first modification involves transposing the structure of the tbl object. This transposition structurally changes the tbl object consistent with the Mosaic Plot's visual layout. Introducing a new R function called *t*, the following code is used to execute this transposition:

```
t(tbl)
```

```
          x
y           01-03 04-12 13-36 37-84 85+
    500       177   231   400   526 322
    501-800    60   113   174   234 163
    800         7    12    22    34  25
```

The transposed table shows the proportional values as visually represented in the Mosaic Plot. For example, the value 177 is the total number of customers with a credit score of 500 or less. It is also the number of customers who have maintained an account with ACME Bank for less than 3 months. On the opposite end of the table, there are 25 customers with a credit score above 800 who have also maintained an account with ACME Bank for more than 7 years.

The second modification required adds column and row sum totals, called margins, to the tbl object. To retrieve both column and row totals, a function called *addmargins* is called. The code and its corresponding output are provided below:

```
addmargins(A = t(tbl), margin = 1)

          x
y           01-03 04-12 13-36 37-84 85+
    500       177   231   400   526 322
    501-800    60   113   174   234 163
    800         7    12    22    34  25
    Sum       244   356   596   794 510
```

```
addmargins(A = t(tbl), margin = 2)

          x
y           01-03 04-12 13-36 37-84 85+  Sum
    500       177   231   400   526 322 1656
    501-800    60   113   174   234 163  744
    800         7    12    22    34  25  100
```

Two arguments are used in the *addmargins* function. The first argument, called *A*, references the transposed table object called tbl. The *margin* argument refers to the dimension type covered by the sum-total action. The argument accepts a numeric value of either 1, 2, or both 1 and 2. A value of 1 conducts a sum-total by column, and 2 by row. Alternatively, both column and row summaries can be combined in the margin argument to create one complete data object:

```
addmargins(A = t(tbl), margin = c(1, 2))
              x
y            01-03 04-12 13-36 37-84   85+  Sum
    500        177   231   400   526   322 1656
    501-800     60   113   174   234   163  744
    800          7    12    22    34    25  100
    Sum        244   356   596   794   510 2500
```

Based on the modifications made to the tbl object, a general assessment of this data reveals some interesting facts. There are more ACME bank accounts that have been opened between 3-7 years (794) than any other age group of accounts. In addition, there are more ACME Bank customers with a credit score of 500 or less (1,656) than those within the other two groups combined!

To add another dimension of analysis to the state of ACME Bank's customer base, let's look at percentage as a function of these proportions. To retrieve this information, another function will be introduced called *proportions*. Nesting the *proportions* function within the *t* function used to create a transposed *two-way table* will convert the tbl object's proportions into corresponding percentages:

```
proportions(t(tbl))
             x
y             01-03   04-12   13-36   37-84     85+
    500      0.0708  0.0924  0.1600  0.2104  0.1288
    501-800  0.0240  0.0452  0.0696  0.0936  0.0652
    800      0.0028  0.0048  0.0088  0.0136  0.0100
```

Applying the same code syntax structure used to retrieve previous proportional values and sum totals, a more complete percentage table can be generated:

```
round(x = addmargins(A = proportions(t(tbl)), margin = c(1, 2)), digits = 2)
              x
y            01-03 04-12 13-36 37-84   85+  Sum
    500       0.07  0.09  0.16  0.21  0.13 0.66
    501-800   0.02  0.05  0.07  0.09  0.07 0.30
    800       0.00  0.00  0.01  0.01  0.01 0.04
    Sum       0.10  0.14  0.24  0.32  0.20 1.00
```

To facilitate the readability of the table percentages, each value was rounded up and truncated. As the code syntax shows, a function called *round* was added. This newly introduced function is supported by two arguments, the first of which contains nested functions. The *round* function's first argument is called x, which refers to a numeric vector called tbl. However, in this example, tbl is a modified data object of class table. The second argument, called digits, determines the number of decimal places to use during the rounding process.

Notice the number of percentages, after rounding, that display a value of 0.00. Two of these percentages, naught in value, are found on the 800 row. Zero-based values displayed in a truncated percentage table impede data context. There is a difference between 0 and values less than 1%. When analyzing qualitative data as a function of percentage, it is important to also evaluate corresponding data points in the form of category counts and column-row sum totals.

Up to this point in the exercise, output results have been viewed through the Console window. To save the data results used to create a Mosaic Plot, apply a new function called *as.data.frame.matrix*. By applying the *as.data.frame.matrix* function, the tbl object can be easily converted into two separately defined data objects of class data.frame:

```
propdf = as.data.frame.matrix(addmargins(A = t(tbl), margin = c(1, 2)))

percdf = as.data.frame.matrix(round(x = addmargins(A = proportions(t(tbl)), margin = c(1, 2)), digits = 2))
```

Both of these code examples apply a series of compound nested functions. Ultimately, the code is designed to support a data object conversion. Each version of the *as.data.frame.matrix* function uses a single argument. In the examples provided, the output is formalized by conversion. The conversion creates a proportioned data frame called *propdf* and a percentage data frame called *percdf*.

Alternative Code Construction Method

Code syntax accuracy, functional knowledge, and patience each play a critical role in successfully constructing executable R code. While the concept of a function may be easy to understand, correctly applying it in code can be a daunting task. Compound nested functions, for example, can generate quite complex code syntax. Consequently, some of the code examples provided in this chapter use nested functions. For a beginner, one of the biggest challenges in R is to correctly construct and apply nested functions. Knowing how to build and successfully apply code in this fashion improves code development skills.

However, there is an alternative code development method to the nested function. Decomposing nested functions into smaller code chunks is often easier to develop, apply, and understand. For example, deconstructing the compound nested function used to create the *propdf* object generates the following code:

```
tbl = table(x = bnk$actage, y = bnk$score)

obj1 = t(tbl)

obj2 = addmargins(A = obj1, margin = c(1, 2))

propdf = as.data.frame.matrix(obj2)
```

Deconstructing the compound nested function used to create the *percdf* data frame produces the following code:

```
tbl = table(x = bnk$actage, y = bnk$score)

obj1 = t(tbl)

obj2 = proportions(obj1)

obj3 = addmargins(A = obj2, margin = c(1, 2))

obj4 = round(x = obj3, digits = 2)

percdf = as.data.frame.matrix(obj4)
```

Do you see a pattern? The key to successfully deconstructing R code is to develop small code chunks through which data objects are created then subsequently layered. For example, object 4 builds on object 3 which builds on object 2 which builds on the foundation defined by object 1. This code development method is similar to the concept of plot layering previously described in this chapter on Line Plots. However, this coding method comes with a downside. Deconstructing code can clutter the RStudio workspace with several data objects. If not properly managed, excessive data object clutter can complicate a workflow. If a workflow is disrupted by this complication, productivity will be negatively impacted.

Using code chunks as a means to deconstruct data processing and computation yields the same results as a nested function. Choose the code development method that best suits your workflow style.

The Current State of ACME Bank's Customer Base

With a preliminary data analysis complete, a baseline version of the current state of ACME Bank's customer base is now ready for review. Referring back to ACME Bank's initial meeting, a question was posed. Given the bank's active position in wanting to serve a customer base that exemplifies a high degree of credit worthiness, how many customer accounts would need to be removed from the system to meet this objective? The baseline used to create the Mosaic Plot provides the context needed to continue this discussion:

- The Mosaic Plot compares three groups of customer credit scores against five categories of open and active accounts.

  - Are these measures appropriate for this analysis?
  - Should the customer credit score categories be disaggregated to reflect more or less categories, or are they acceptable in their current form?
  - If the customer credit score groups are to be disaggregated, by what credit score measure will each category be defined?
  - Subsequent to a disaggregation of more credit score categories, how many total categories will there be?
  - Should the open and active account categories be extended, reduced, or remain as currently defined?
  - If the active account categories are to be either extended or reduced, by what account aging measure will each group be categorized?
  - Subsequent to either an increase or a decrease in active account categories, how many total active account categories will there be?

- The baseline sample size represents 2,500 customer accounts.

  - Is 2,500 an appropriate sample size for this analysis?
  - Should a larger sample size be used?
  - If a larger sample size is used, how many customer accounts will define the new sample size?

These are some of the questions that the Mosaic Plot provided in this exercise can answer. In a typical business environment, the baseline would be refined and a new analysis would be conducted. An updated Mosaic Plot would then be generated, after which discussions would continue. For example, ACME Bank may decide to refine the analysis by including additional measures. In addition to customer credit scores and aged account information, ACME Bank may also want to know what type of account each customer holds. Is it a savings, checking, business, or a trust account? ACME Bank may also want to know how many active accounts each customer has. While this is a quantitative measure, it can be qualitatively converted. For example, irrespective of the number of active accounts a customer may have, the results can be grouped into one of three categories. The customer may have one, two, or more than two accounts.

Conclusion

More than two categorical variables can be used to create a Mosaic Plot. However, adding data variables to a Mosaic Plot increases its visual complexity, making it more difficult to interpret. To limit a Mosaic Plot's visual complexity, it is recommended that no more than three categorical variables be used in its design. For the beginner using R, designing a Mosaic Plot with two categorical variables keeps the code simple and generates a powerful plot that is easy to explain and understand.

Scatter Plot

Where the Mosaic Plot is qualitative by design, the Scatter Plot is quantitative in form. In fact, quantitative data represent the centerpiece of a Scatter Plot. As the name implies, a Scatter Plot uses numeric values as points that are distributed or "scattered" across a coordinate grid defined by an x and y axis. The result returns a plot design in which patterns, clusters, correlations, and outliers can be visually identified. If constructed correctly, the Scatter Plot can be used as a compelling tool in developing a data narrative.

The most interesting insights derived from a Scatter Plot often emerge when plot points are disaggregated into groups called clusters. In a Scatter Plot, clusters are typically distinguished by color.

In this exercise, the performance measurements of a manufacturing process taken each day over a three-month period will be evaluated. The objective of this evaluation is to build a data narrative that achieves the following:

- Establish an operational baseline for manufacturing process performance
- Determine conclusively whether the manufacturing process is performing at, above, or below its operational capabilities

Simulating A Dataset

The dataset used in this exercise simulates the performance metrics of a manufacturing process. The R code used to generate this dataset, called *manuf_proc*, is provided below:

```
set.seed(seed = 1999, kind = "Mersenne-Twister")

meas = abs(rnorm(n = 90))

mth = as.factor(append(x = rep(x = "Apr20", times = 30), values = c(rep(x = "May20", times = 30), rep(x = "Jun20", times = 30))))

seg = append(x = rep(x = "red", times = 30), values = c(rep(x = "green4", times = 30), rep(x = "orange", times = 30)))

manuf_proc = data.frame(meas,  mth, seg)
```

```
manuf_proc$mnstat = ifelse(manuf_proc$meas > mean(manuf_proc$meas), "ABV", "BLW")

manuf_proc$mnstat = as.factor(manuf_proc$mnstat)
```

To review the first six records of the *manuf_proc* dataset, use the head function:

```
head(manuf_proc)
       meas   mth seg mnstat
1 0.73267249 Apr20 red    BLW
2 0.03782971 Apr20 red    BLW
3 1.20300914 Apr20 red    ABV
4 1.46980203 Apr20 red    ABV
5 0.13369030 Apr20 red    BLW
6 0.51982725 Apr20 red    BLW
```

A brief description of each vector comprising the *manuf_proc* dataset is provided as follows:

- **meas** – A numeric value that represents a performance measurement from the manufacturing process. One measurement is taken towards the end of each day. The performance values are created by simulating a normal distribution of data.

- **mth** – Defines the month and year in which the performance measurement was taken. Months simulated for this vector are April, May, and June of 2020.

- **seg** – This value determines the cluster segment. Each cluster segment represents a period of one month consisting of 30 consecutive days. A total of three cluster segments are represented. Segments are differentiated by three color categories. Colors used are red, green, and orange.

- **mnstat** – This metric represents the Mean Statistic relative to all performance measurements taken. This data point returns either *ABV* for above, or *BLW* for below relative to the dataset's performance measurement mean.

The dataset row identifier, while not defined as an actual data field, plays a critical role in providing context to the *manuf_proc* dataset. Each row identifier represents a day number on which a performance measurement was taken. There are 90 records comprising the dataset.

The code that generates the *manuf_proc* dataset departs slightly from the code used to simulate data in previous exercises. As a result, a brief explanation of each code line supporting the creation of the *manuf_proc* dataset is provided below:

- **set.seed** – A function used to generate data reproducibility, preserving both the form and order of data. This function uses a numeric argument called *seed*. The seed value represents the function's starting point for generating pseudo-random data.

- **meas** – This numeric vector is created by applying two critical functions:

  - abs – This function ensures that each measurement is an absolute value. This means that all numeric values passed through this function return a positive number.
  - rnorm – This function generates a normal distribution of 90 numeric values. These values, defined by the argument *n*, are computed by an approximation of the mean and standard deviation of the distribution. Both operations

represent arguments in the function. On this code line, the *rnorm* function applies the default values of 0 for the mean and 1 for the standard deviation, implicitly defined.

- **mth** – This vector of class factor introduces two new functions used to create the data:

  - append – This function, which applies two primary arguments, *x* and *values*, adds elements or groups of elements to a vector. In this code example, three groups of elements create the mth vector. The *x* argument defines the first segment group of 30 elements. The *values* argument specifies then subsequently combines the second and third groups of 30 elements to the mth vector for a total of 90 elements.

  - rep – This function repeats a specified action. In the nested function used to create the mth vector, rep generates 30 duplicated elements for each of the words "Apr20," "May20," and "Jun20." The rep function itself is executed three times within this code line after which the results are appended together. The argument *x* may create some initial confusion as it represents a duplicated argument name within two different functions. To clarify, the append *x* argument applies to the grouped vectors being combined while rep *x* defines the elements comprising each group.

- **seg** – Using color names instead of months, this code line applies the same code syntax pattern as provided by the mth code line. The only syntactic difference in the code is that the vector output was not converted to class factor.

- **manuf_proc** – This code line generates the data frame object used in this exercise by combining the meas, mth, and seg vectors.

- **manuf_proc$mnstat** – This code line creates a new data vector that is added to the *manuf_proc* dataset. A binary logic test is conducted to compute the mean of the meas vector. Once identified, the test then determines whether each of the meas values is above or below the mean. If the meas value is above the mean, the status returns ABV. If the meas value is below the mean, BLW is returned.

- The final code line converts the mnstat vector from class character to class factor. In conjunction with the first principles of R, the mnstat vector is qualitative in form, justifying the data type conversion. While the seg vector also meets the criteria as a qualitative measure, it was not converted to class factor. This is because in some cases during a data modeling process, more than one data type conversion is required for a data vector to be effectively used. For example, if the seg vector is used in a data summary analysis in which the *summary* function is called, it would need to be encoded as class factor. Conversely, if the seg vector is used as the color argument in a plot, it would need to be encoded as class character. One way to mitigate this data type conflict is to make a copy of the seg vector, then separately re-encode its data type. For example, the following code will copy, rename, then re-encode the seg vector as a new vector of class factor:

```
seg_fctr = as.factor (manuf_proc$seg)
```

In this example, the new vector, called seg_fctr, can now be successfully used in a *summary* function. In addition, the seg vector, in its current form, could be passed to a plot's color argument.

It is important to be mindful of the nuances in code syntax and function. These nuances will impact the R user experience. Nuances in code syntax influence whether code can be successfully executed and whether data as a plot can be successfully generated.

Scatter Plot Components

Once a dataset has been defined and a plot has been selected, the process of developing and modeling code begins. For this exercise, the code needed to generate a baseline Scatter Plot is provided below:

```
plot(x = manuf_proc$meas, xlab = "Day Nbr", ylab = "Measure", pch = 16, col = manuf_proc$seg, las = 1,
main = "Manufacturing Process Performance by Scatter Plot (n=90)")

text(x = manuf_proc$meas, labels = rownames(manuf_proc), pos = 2, cex = .8, col = "black")

mn = mean(manuf_proc$meas)

subtxt = paste0("(mean=", round(x = mn, digits = 2), " | abv=38 | blw=52)")

mthnm = as.character(unique(manuf_proc$mth))

legend(x = "topleft", legend = mthnm,  title = "Mth", bg = "gray88", fill = c("red", "green4", "orange"), cex
= .7)

mtext(subtxt)

abline(h = mn, col = "blue")

grid()
```

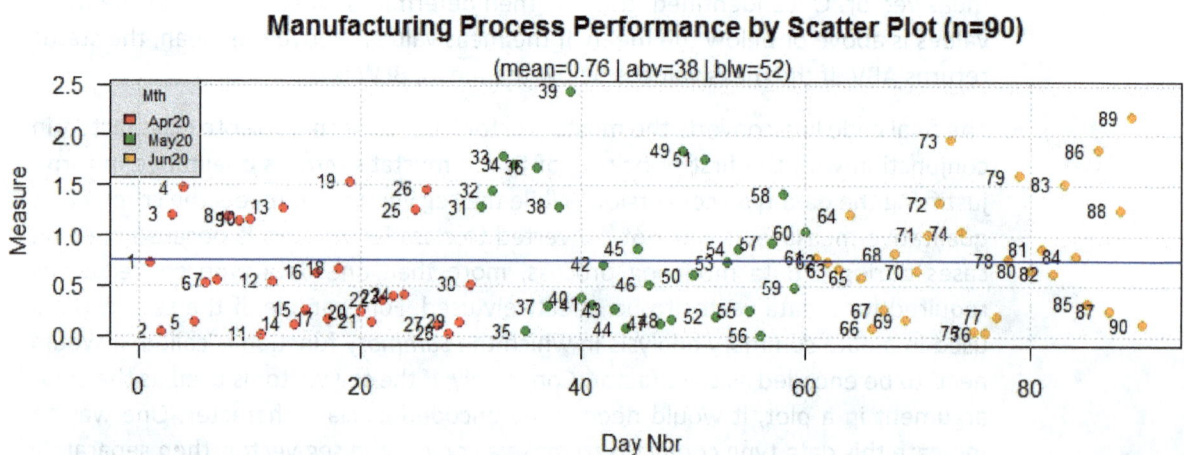

Manufacturing Process Performance by Scatter Plot (n=90)
(mean=0.76 | abv=38 | blw=52)

Upon the execution of the last line of this code script, a Scatter Plot will be generated. The code syntax used to create the Scatter Plot is similar to that provided in previous exercises. However, there are five new functions that have been introduced into this code script that require additional explanation. Additional remarks are also provided to facilitate a broader understanding of code construction and syntax.

Each line of code should be viewed as a "layer" upon which subsequent code functionality is superimposed.

- **1ˢᵗ Code Line:** This code line establishes the plot's parametric foundation. It includes the vectors to be plotted, label names, marker size and color, axes tick mark direction, and the plot's main title. In this Scatter Plot example, the row numbers in the *manuf_proc* dataset define, by implication, the plot's y axis.

- **2ⁿᵈ Code Line:** This code line adds marker labels to the Scatter Plot. The labels argument introduces a new function called *rownames*. The *rownames* function directly references the data points that define the plot's y axis. In addition, this code line sets marker position, color, and size.

- **3ʳᵈ Code Line:** Computes the mean of the measurements defined by the *manuf_proc* dataset's meas vector. The results are saved as a separate data object called mn.

- **4ᵗʰ Code Line:** This code line creates the text vector used in the Scatter Plot's subtitle. The text vector name is subtxt. The *paste* function was previously introduced in this chapter to create Fan Plot labels. However, a variation of this function, called *paste0* can also be used to build plot text content. The difference between the *paste* and *paste0* functions largely involves how the spacing between words is controlled. The *paste0* function generally offers more control over spacing than its functional counterpart. However, both functions are invaluable tools used to effectively develop plot labeling and title content. In this Scatter Plot example, three data points are used to create the plot's subtitle:

  - The meas vector's mean
  - The total number of measurements above the mean value
  - The total number of measurements below the mean value

The *round* function is used to truncate the mean value to two digits. The pipe, a symbol represented by a vertical line, separates each data point. Open and closed parentheses are used to visually differentiate the plot's subtitle from its main title. The totals represented by *abv* and *blw* can be extracted by applying the *summary* function:

summary(manuf_proc$mnstat)

ABV   BLW
38   52

A plot subtitle is quite different from its main title. A statistically-driven plot's main title defines the plot in the most general terms. Conversely, a plot's subtitle adds important context to a plot. In addition, plot subtitles provide critical information used to compliment plot interpretability.

- **5ᵗʰ Code Line:** This code line generates a vector called mthnm. It is of class character. It captures the segment names used in the Scatter Plot's legend. It applies a new function called *unique*. The *unique* function extracts all unique values from a vector. In this example, the mth vector is copied, parsed, and converted from class factor to class character. A data type conversion of the mthnm vector is not required for this code script to correctly execute. However, it is good practice for the textual attributes of a plot to be of class character. This practice creates continuity in the construction and management of vector data types.

- **6th Code Line:** The sixth code line generates the Scatter Plot's legend. The *legend* function sets the legend object's position, size, content, title, background color, and fill color.

- **7th Code Line:** This code line introduces a new function, called *mtext*. It is used as a plot-building function in R, providing an efficient way to add a subtitle to the upper margin of a plot. While the *mtext* function contains several arguments, the principal argument used is the subtitle content itself. In this example, the *mtext* function uses the subtxt vector created in the fourth code line.

  An argument called side can be added to control subtitle placement. The default placement of the subtitle is the upper margin of the plot but three other placement options are available. Plot margins are numerically defined. The side argument accepts a numeric value from 1 to 4. Three is the numeric default.

- **8th Code Line:** A newly introduced function called *abline* generates the blue line that runs horizontally through the Scatter Plot. The *abline* function is easy to use. It can be applied by using a few principal arguments. These arguments are briefly described below:

  - *h* – This argument determines both the line direction and its placement on a plot. The line drawn by this argument will be horizontal in form.

  - *v* – This argument determines both the line direction and its placement on a plot. The line drawn by this argument will be vertical in form.

  - *col* – This argument sets the line color.

  A line is added to a plot to provide context to a specific data attribute or to accentuate an aspect unique to a plot's data narrative.

- **9th Code Line:** This code line finalizes the Scatter Plot by adding a light, gray-colored background grid. The grid color can be changed by adding a color argument to the *grid* function called *col*. Additional arguments can be used to set the grid object's line type and width.

Evaluating A Scatter Plot

The Scatter Plot for this exercise has been successfully created. What does the data narrative convey? Does the Scatter Plot's data narrative satisfy the evaluation objectives posed by this exercise? An examination of the data narrative reveals several insights to consider.

Perhaps the most obvious fact supporting the data narrative is the manufacturing process performance baseline. The blue line that runs through the length of the plot is positioned on the y axis at precisely .759861. This line establishes the operational baseline for manufacturing process performance. More specifically, the third code line of the script captures this measurement and code line eight plots it.

Understanding the relationship between the operational baseline and the general time period of manufacturing process performance is another critical part of the data narrative. Relative to this relationship, the Scatter Plot shows that the operational performance of the manufacturing process recorded more measurements below its operational performance mean than above it. The plot's subtitle shows that over a three-month period, 52 process measurements were recorded below the performance mean, and 38 were above it. Translated, this means that approximately 58% of all performance measurements recorded during this period (52÷90) returned a performance level generally below the system's operational performance mean.

Exploring the data narrative further, the temporal segments of the system's manufacturing process performance can be examined. The *range* function extracts both the minimum and maximum values from a numeric vector. By applying the *range* function to the meas vector, minimum and maximum performance measurement values can be identified. Meas vector measurements range on the low end from 0.01088569 to a high of 2.43772846.

range(manuf_proc$meas)

[1] 0.01088569   2.43772846

A quick sort of the meas field reveals that the lowest process performance measurement was captured on day number 11 in the Apr20 temporal segment. Conversely, the highest performance measurement was recorded 28 days later on day number 39 in the May20 temporal segment.

Another metric that can be examined is the performance variance for each temporal segment. This metric measures the distance of temporal segment performance values against the mean. Each variance measures the proximity spread of a temporal segment's performance measurements. The idea is to identify the consistency of the system's performance by temporal segment.

The smaller the variance, the higher the degree of closeness that exists between performance measurements. A smaller variance is indicative of a more consistently performant system. However, a higher variance represents a more dispersed proximity spread of measurements, suggesting that the system is less consistent in its performance. A key element in this data narrative is not only to find each segment's variance, but also to understand the system's operational impact relative to its performance, above or below the baseline mean.

Variances for each temporal segment can be calculated by applying the *aggregate* function.

aggregate(formula = meas ~ mth, data = manuf_proc, FUN = "var")

```
  mth      meas
1 Apr20    0.2523981
2 Jun20    0.3497942
3 May20    0.4477031
```

The final metric used to develop the data narrative involves the acquisition of performance measurements by group. This metric computes the total number of performance measurements that exist both above and below the mean for each temporal segment. To retrieve this information, a two-way table can be used:

table(manuf_proc$seg, manuf_proc$mnstat)

```
         ABV    BLW
green4    14     16
orange    15     15
red        9     21
```

To better understand the analytical context defining the data narrative, three critical metrics have been identified. These metrics are as follows:

- The system's 3-month performance measurement range
- Temporal segment variances
- Temporal segment measurement counts above and below the performance mean

Consistent with the development of a Scatter Plot, and given these metrics, what conclusions can be drawn as a means by which to shape the data narrative? Before any conclusions are drawn, facts supporting the data narrative must first be stated. The data metrics captured in this exercise can be used to weave a data narrative that supports the following facts:

- The system's manufacturing process performance baseline is .759861 with measurements ranging from 0.01088569 to 2.43772846.

- The temporal segment with the smallest performance measurement variance was Apr20; the largest variance was May20.

- 58% of all performance measurements recorded during the three month assessment period returned a performance level generally below the system's operational performance mean.

- The temporal segment with the smallest variance also contained the smallest performance measurement in the measurement range (Apr20).

- The temporal segment with the largest variance also contained the largest performance measurement in the measurement range (May20).

- The temporal segment with the smallest variance also contained the most number of measurements below the system's performance mean (Apr20).

- The Jun20 temporal segment recorded the same number of performance measurements both above and below the system's performance mean baseline.

Conclusions

A primary characteristic of a data narrative is that both facts and conclusions are structured in a way that can easily be used in a presentation. Facts and conclusions, as component parts of a data narrative, should support the plot visualization(s) being used. Facts derived from data should also be thematically related to plot elements, ensuring that an unequivocal connection exists between explanation and interpretation.

Based on a set of statistically driven facts, what final conclusions can be drawn? Ultimately, conclusions derived from a data analysis should *solve for x*. In its simplest terms, *solving for x* means finding the truth[19] to a question being posed. In this example, *solving for x* means achieving the objectives posed by this exercise. Pursuant to these objectives, the following conclusions are drawn:

- The system's operational baseline for manufacturing process performance is .759861.

- The system's manufacturing process generally operates below its performance baseline

The conclusions provided by this exercise set an important precedent. In many ways, the conclusions generate more questions than answers. For example, is it a positive or a negative indicator that the system's manufacturing process operates below its performance baseline? Is the .759861 performance baseline consistent with performance baselines captured in the past? How will this data compare to similarly captured data in the future? How do the temporal segment variances compare with historical temporal segments? Are there any variance patterns that emerge? If so, how are the patterns manifested? Are variance patterns more cyclic or seasonal? Are there any trends?

Perhaps the most important question that remains to be answered from the conclusions reached is why?

Final Remarks

With a dataset consisting of only four data fields and 90 records, a very small part of a larger data analysis was conducted. This exercise demonstrates how a single plot can be used to construct a data narrative that *solved for x*. Other plots could have been used to provide additional context to the data narrative including Box Plots, Histograms, and a host of other plots specialized for use with quantitative data.

Due to the sheer number of ways in which data can be statistically analyzed today, it is important to control the scope of the data narrative. If the data narrative is too broad, it will be interpreted as excessively confusing and complex. If it is too narrow, the data narrative will appear implausible by virtue of simplism. However, if the data narrative is constructed correctly, it should be accepted as both credible and persuasive.

---

[19] In many cases, *solving for x* means finding the highest probability of truth for a question posed. For example, the statement, "there is an 85% chance that the baseline will remain less than .76 for the next 8 weeks" *solves for x* using a probability of truth. In this manner, *solving for x* is predictive in form. *Solving for x* can also be framed in binary terms of likelihood defined by the phrase, "more or less likely."

NOTES

CHAPTER 5:
APPLICATION OF COLOR IN PLOT DEVELOPMENT

One of the most important aspects of plot development in R after plot type selection is defining the plot's color scheme. Color is fundamental to a plot's design and selecting the right color scheme remains a critical part of plot development. Correctly designed, colorized plots can be exceptionally persuasive and compelling in describing a data narrative. In addition, color schemes that maintain the right balance of luminance and contrast add texture and style to a plot. Relative to aesthetic dynamics, color over gray-scaled plots are generally preferred. While adding color to a plot in R is not a difficult task, the fundamental challenges in applying color in R are as follows:

- Knowing which functions and packages to use from the sheer raft of options available

- Knowing the circumstances under which color combinations are applied

Understanding the context in which color is used in plot development requires a familiarity with the *first principles of plot colorization*. Similar in construction to the First Principles of R, the first principles in the application of color in plot development can be distilled into three fundamental precepts. These precepts, along with visual examples, are defined in the table below:

First Principles of Plot Colorization in R

FIRST PRINCIPLE	DESIGN THEME	EXAMPLE
Understanding color in it is application as a unidimensional theme	As a monochromatic color	
Understanding color in it is application as a palette	As a collection of non-gradient colors	
Understanding color in it is application as a gradient	Color as a manifestation of shades	

This chapter provides a fundamental overview of color application in R. *Many details are left out by design, exchanging an exploration of depth for a manageable view defined by simplicity. Various references to color-based packages are provided which can be pursued further, if a curiosity for a deeper understanding of color application in R remains a continued interest.*

Understanding Color in its Application as a Unidimensional Theme

The most rudimentary application of color in a plot occurs when one color can adequately define the plot's primary graphical elements. A plot generated with one color, called a Monochromatic Plot, is typically used to present a unidimensional theme such as a single cluster in a Scatter Plot, a small number of categories in a Bar Chart or Histogram, or where categorical comparisons show nominal changes in feature similarity.

Example 1:
```
set.seed(seed = 109, kind = "Mersenne-Twister")

ds = abs(rnorm(100))

plot(x = ds, col = "blue", pch = 16, ylab = "Val", main = "Scatter Plot With Unidimensional Color Theme")
```

Example 2:
```
data("trees")

hist(x = trees$Height, xlab = "Tree Height", main = "Monochromatic Frequency-Based Histogram", col =
"orchid",  las = 1)
```

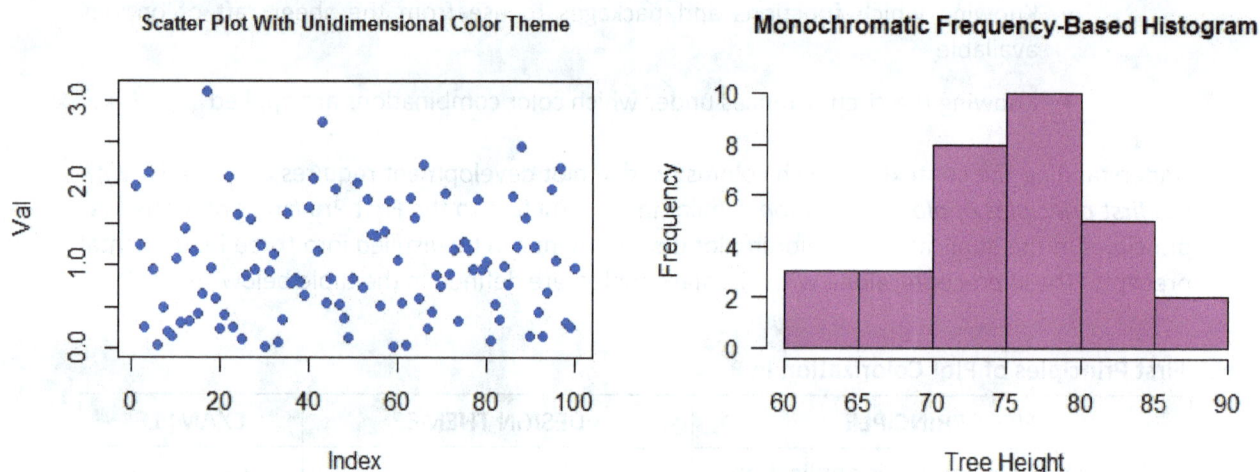

Scatter Plot With Unidimensional Color Theme

Monochromatic Frequency-Based Histogram

Understanding Color in its Application as a Palette

Plot development in which non-gradient colors are used to construct a palette provide the polychromatic framework on which a graphical representation of data is built. Unlike the application of color as a unidimensional theme, a Polychromatic Plot uses color to add nuance and specificity to its presentation. In addition, using a polychromatic palette as a plot's color scheme is extremely effective in highlighting one or more characteristics within a data narrative. Various packages in R facilitate the development of color schemes in which color as a palette can be used to create compelling Polychromatic Plots. There are two ways to create a Polychromatic plot using color as a palette:

- As a user-defined color scheme

- As a pre-defined color scheme

The Palette as a User-Defined Color Scheme

With maximum ease and flexibility, a user-defined color palette provides the means to aggregate any combination of colors to build a plot palette. The only drawback to this approach is that the combination of colors selected may not meet the luminance and contrast standards typically required to generate a visually distinctive color palette. To design a user-defined color palette, simply create a vector of class character with a listing of R-compatible color names. Consideration should be taken to determine both the sequential order and the total number of color names used to create the color palette. Once the palette has been created, it is then directly referenced in the color argument of a plot function. As of this writing, there are a total of 657 R-compatible color names available for use in R. The following code returns R's color directory through a vector of class character:

```
col = colors()
```

In the following plot example, the color palette variable called *cpal* represents the user-defined color scheme. The key to successfully deploying a user-defined color scheme is to ensure that the number of palette colors precisely match the total number of bins or categories in the plot.

```
data("trees")

cpal = c("red", "green4", "blue", "orange", "black", "yellow")

hist(x = trees$Height, main = "Frequency-Based Plot with a User-Defined Color Scheme", col = cpal,
xlab = "Tree Height", las = 1)
```

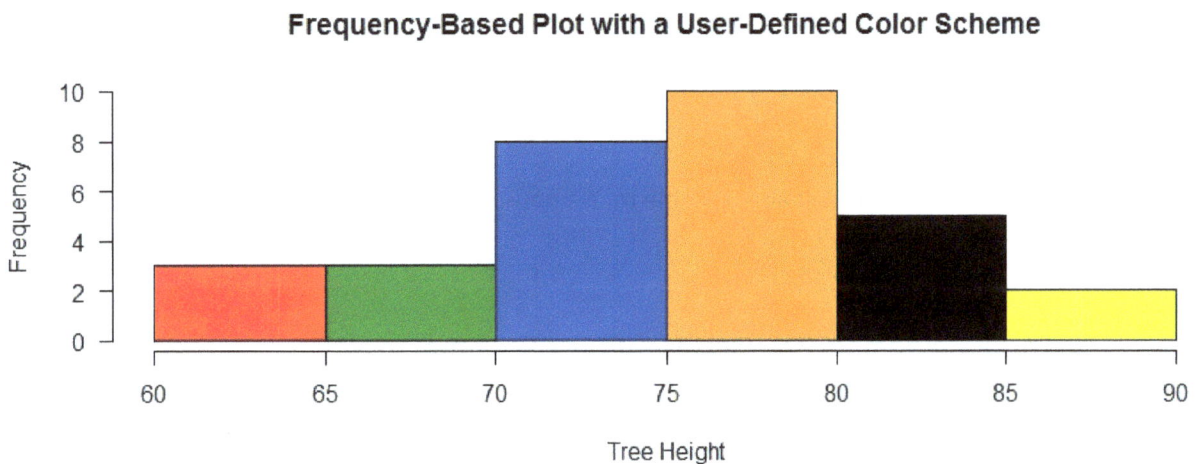

To retrieve the total number of bins in a Histogram, apply the following code:

```
hobj = hist(x = trees$Height, plot = FALSE)

length(hobj$counts)
```

[1] 6

In concluding a summary explanation of user-defined color schemes, it is important to recognize a unique color feature in R. There is a built-in capability that connects numeric values to color names. These numeric values are referred to as *color constants*. Color constants range in value from one to eight. This means that instead of color names, numeric values can be used in a plot's color argument. These constants and their related color keys are displayed in the table below:

R Color Constant Table	
NUMERIC VALUE	COLOR EQUIVALENT
1	BLACK
2	RED
3	GREEN
4	BLUE
5	CYAN
6	MAGENTA
7	ORANGE
8	GRAY

This R color feature can be used to facilitate accelerated plot development when color is a secondary consideration. For example, if a dataset contains numeric values in a particular vector ranging from one to four, a plot could be categorically represented by connecting the numeric values to R's built-in color scheme. If numeric values are used, no more than eight categories within a vector should be represented. To provide an example, modify the *cpal* variable used in the last exercise then re-execute the code:

cpal = 1:6

Using this method, it is also possible to use non-contiguous numbers to control for color:

cpal = c(1, 4, 3, 8, 6, 2)

The Palette as a Pre-Defined Color Scheme

A pre-defined color scheme in R uses a pre-configured color palette. The user does not choose the colors but instead selects the palette. Creating a Polychromatic Plot by applying a pre-configured color palette meets two critical requirements for optimizing plot colorization in R. Pre-defined color schemes in R:

- Provide an optimal balance between luminance and contrast

- Are adequately separated within the color space

Three pre-configured color palettes are strongly recommended for use by beginning users of R in that the package from which they derive is already loaded. Found in the *grDevices* package, these palette functions are listed as follows:

rainbow(12)

terrain.colors(12)

topo.colors(12)

The palette functions' numeric value represents a reasonable upper bounds of unique colors found within the color palette. It is possible to use a numeric value for a palette that is either smaller or larger than the upper range recommended for a plot. However, if applied in this manner, the plot's color results will vary. For example, if the numeric value is smaller than the total number of categories defined by a plot, the colors will repeat. If the numeric value is larger than the total number of categories, the results will yield a more gradient output. For optimum color visualization, the palette function's numeric value should match the total number of categories defined by the plot.

An obscure characteristic of pre-defined color palettes in R is that most of them return color values expressed as *Hex Color Codes*. A Hex Color Code consists of a hash followed by a six-digit alphanumeric combination of values designed to represent variants of red, green, and blue. For example, extracting five colors from the topo.colors palette generates five Hex Color Codes:

```
topo.colors(5)
[1] "#4C00FF"    "#004CFF"    "#00E5FF"    "#00FF4D"    "#FFFF00"
```

For the user, this output provides indiscernible color information related to the topo.color palette. To convert Hex Color Codes to a readable color format, a function provided in the DescTools R package called *HexToCol* can be used. It converts Hex Color Codes to the closest color match available in R.

```
library(DescTools)
HexToCol(topo.colors(5))
[1] "blue"     "dodgerblue2"     "turquoise1"     "springgreen2"     "yellow"
```

The code example below generates a Bar Plot of automobile types and counts from a dataset called *Cars93*. Using the same data, configuring the plot with different color palettes by modifying the *cpal* variable generates a different color experience. The *rainbow* palette reflects a primary-based color scheme, the *terrain.colors* palette is more earthy in color, and *topo.colors* is more topological in tone, deriving its color palette from cartography.

```
install.packages("MASS")

library(MASS)

data("Cars93")

cpal = rainbow(6)

plot(x = Cars93$Type, xlab = "Auto Size Type", col = cpal, ylab = "Count", main = "Bar Plot With Pre-Defined Color Scheme\n(rainbow)")
```

cpal = rainbow(6) cpal = terrain.colors(6) cpal = topo.colors(6)

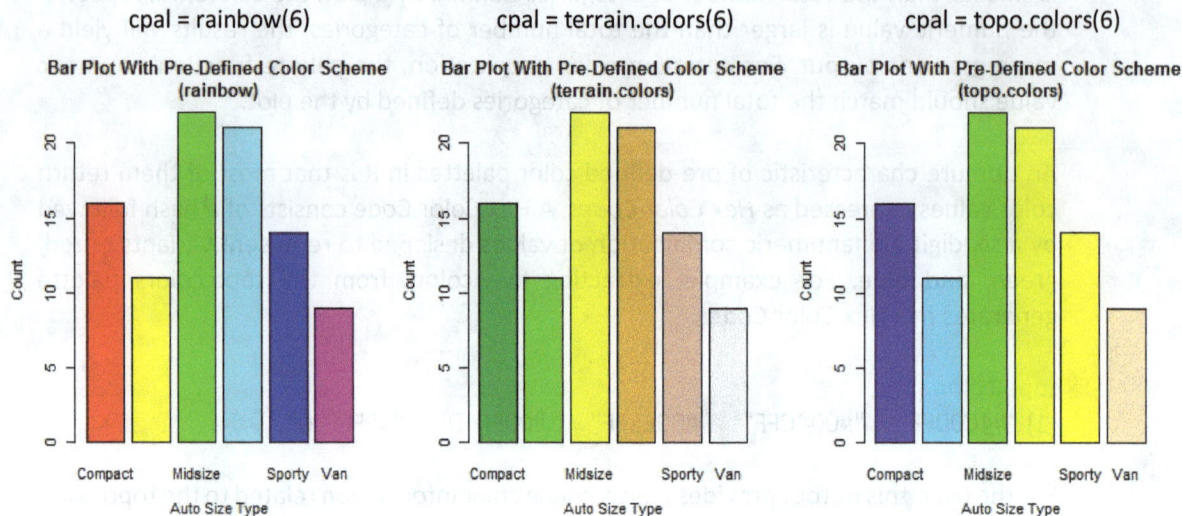

Notice the formatting protocol provided in the plot's *main* argument. The "\n" symbol generates a hard line break in the plot's title, controlling for data parsing between lines.

In addition to the color palettes previously mentioned, the table below lists 18 of the most compelling non-gradient, pre-defined color palettes available in R along with their associated package name. While the list is by no means exhaustive, it provides a balance between manageability and utility. Each color palette contains at least 10 uniquely defined colors. In addition to each palette's remarkable color layout, these color palettes are recommended due to the ease with which they can be applied.

Pre-Defined Color Palettes Used in R

Color Palette Defined by Recommended Upper Bounds	R Package
primary.colors(10)	colorRamps
alphabet(26)	pals
alphabet2(26)	pals
cols25(25)	pals
tableau20(20)	pals
watlington(16)	pals
gg.col(100)	plot3D
gg2.col(100)	plot3D
jet.col(100)	plot3D

Color Palette Defined by Recommended Upper Bounds	R Package
jet2.col(100)	plot3D
alphabet.colors(26)	Polychrome
dark.colors(24)	Polychrome
glasbey.colors(32)	Polychrome
green.armytage.colors(26)	Polychrome
kelly.colors(22)	Polychrome
light.colors(24)	Polychrome
palette36.colors(36)	Polychrome
sky.colors(24)	Polychrome

Understanding Color in its Application as a Gradient

Gradient colors used in plots are designed to draw attention to a specific theme advocated by the data narrative. This theme is defined by showing a change in magnitude of some physical quantity or dimension demonstrated by the plot. Gradient colorization compliments the theme by generating a color palette defined within a specific color range. The color range is typically represented by a shade of colors extracted from a base or primary color.

In the code example provided below, a gradient Bar Plot shows the color nuances between the height and frequency relationship of Pine Trees taken from the *Loblolly* dataset:

```
data("Loblolly")

install.packages("scico")

library(scico)

cpal = scico(n = 15, palette = "vik")

hist(x = Loblolly$height, col = cpal, xlab = "Height", main = "Pine Tree Height Histogram by Gradient")
```

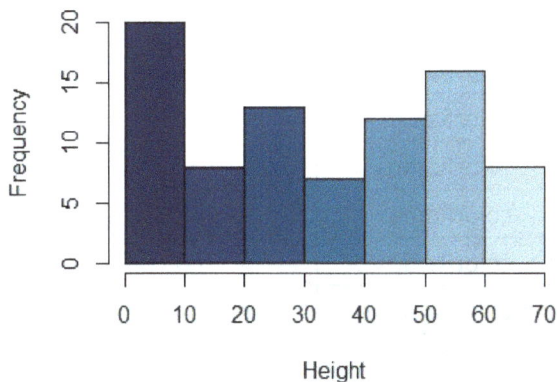

Pine Tree Height Histogram by Gradient

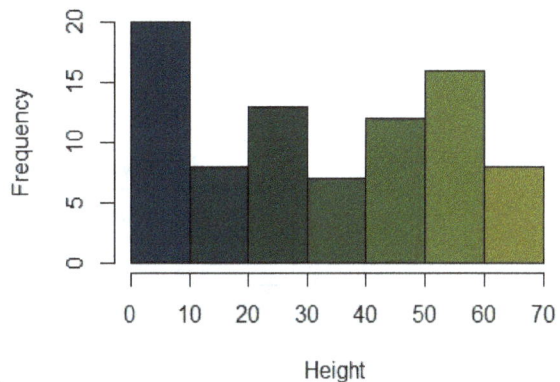

Pine Tree Height Histogram by Gradient

The gradient palettes used to generate these plots were created with the help of the *scico* package. Generating a gradient palette from the scico package requires the *scico* function. In this package, gradient palettes aren't directly accessed, but are first created before used in a plot. To create a gradient color palette, the *scico* function applies two primary arguments, briefly outlined below:

- **n:** A numeric value that determines the number of colors that will be used to create the palette.

- **palette:** The name of the palette framework from which to sample. The palette name used must be enclosed with quotation marks. There are a total of 28 palette frameworks from which to choose. To retrieve a listing of available palette names used in this functional argument, apply the following code:

  scico_palette_names()

  To get a master view of the 28 palette frameworks, use the following code:

  scico_palette_show()

The *cpal* variable used to generate the gradient palette used in the second plot is provided as follows:

cpal = scico(n = 10, palette = "bamako")

The upper bound for a gradient palette is largely determined by the number of "gradients" being plotted. Experimenting with plot-color variations will yield the best results in being able to establish the upper bounds of a gradient palette.

In addition to the gradient palettes previously mentioned, the table below provides 10 additional functions that can be used to generate gradient-based palettes.

Gradient-Based Palette Functions

Function Used to Create Gradient-Based Palette	R Package
blue2green	colorRamps
blue2green2red	colorRamps
blue2red	colorRamps
blue2yellow	colorRamps
cyan2yellow	colorRamps
green2red	colorRamps
magenta2green	colorRamps
ygobb	colorRamps

Function Used to Create Gradient-Based Palette	R Package
cm.colors	grDevices
heat.colors	grDevices

<u>NOTES</u>

CHAPTER 6:
VIDEO LIBRARY OF COMMON TASKS IN R

This chapter outlines how coding algorithms can be used to perform common tasks in R. Previous chapters explored the functional characteristics of code and the means by which to programmatically design and analyze plots. However, a shift in perspective is required to expand the parameters of understanding from code-as-function to code-as-task. Viewing R code through this perspective will help expand the depth of knowledge in understanding the nuances of R technology as it is applied in different ways.

This chapter facilitates this understanding by providing a series of six QR codes, each of which is connected to a video tutorial describing a unique R task. Choose a QR code of interest then open your smart phone's camera application. Scan your smart phone's camera over the code then sit back and watch the tutorial!

If you are unable to successfully access the video library through your phone, go to ***www.youtube.com*** and type the phrase, "Conquering R Basics" in the search box. The search results will return a direct link to the video library. Alternatively, the video library playlist can be accessed through the following link:

https://www.youtube.com/playlist?list=PLrf94eEGvVuE7lnbgBC9C9yXA6EGD3HLq

A brief description of each video tutorial, along with its QR code, is provided below.

TASK 1: Get a Listing of Datasets within an R Package

It is important to be able to efficiently access datasets in R. Datasets provide the foundation upon which an R programming skill set is built. In this video tutorial, you will learn how to extract a listing of datasets provided in an R package. The code is easy to understand and even easier to use.

TASK 2: Convert Various Date Formats to Class Date

Converting date formats compatible with R's class *Date* are problematic for many users, especially when working with imported datasets. There is a way to resolve date-formatting problems in R. In this video tutorial, you will learn how to successfully convert dates into an organized class in R – no manual effort required.

TASK 3: Identify Duplicate Records in a Dataset

Duplicate records pose a real problem in a dataset, compromising the ability to acquire accurate data metrics. Consequently, inaccurate data metrics lead to data misinterpretation. In this video tutorial, you will learn how to effectively identify whether a dataset contains duplicate records. Additional tips are provided to help process duplicate records.

TASK 4: Compute Missing (NA) Data Metrics

Missing data in a dataset impedes the ability to conduct an appropriate data analysis. Unfortunately, in today's world, missing data is a fundamental part of many of the datasets actively being used to process information. There are many ways to handle missing data. This video tutorial will show you how to use R to identify, process, and manage missing data when discovered in a dataset.

TASK 5: Import & Export Datasets

Knowing how to import and export data in R marks an important step in skill set development. A data analysis begins with a dataset that is either directly accessed, programmatically constructed, externally linked, or made available through importation. In this video tutorial, you will be introduced to a set of coding strategies that can be used to successfully import and export data in commonly used file formats, including Microsoft Excel.

TASK 6: Convert a Plot into an Image File (PNG)

With a few mouse clicks in RStudio, a plot generated in R can easily be copied or saved. However, the skill sets needed to programmatically achieve the same result offers a different challenge. In this video tutorial, a three-step process provides a structured way in which R code can be used to convert a plot into an image file. Suitable for inclusion in web pages, databases, spreadsheets, and digital documents, this conversion process creates an image file referred to as a *PNG*, or a *Portable Network Graphics* file.

NOTES